Exercises for the Molecular Biology Laboratory

Dr. Patrick Guilfoile
Biology Department
Bemidji State University, Bemidji, MN

Morton Publishing Company

925 W. Kenyon Ave., Unit 12
Englewood, Colorado 80110

http://www.morton-pub.com

Printed in the United States of America
by Morton Publishing Company
925 W. Kenyon Ave., Unit 12, Englewood, CO 80110

10   9   8   7   6   5   4   3   2   1

ISBN: 0-89582-514-7

Preface

The exercises in this manual are designed to teach you fundamental techniques of molecular biology. These exercises focus on the manipulation and analysis of DNA, RNA, and proteins. Overall, these laboratories will provide you with a solid overview of the core of molecular biology and molecular genetics.

Careful attention to detail and proper procedure is necessary for successful completion of these laboratories. Many of these experiments require measuring and manipulating minute volumes of reagents. An error in measuring of one-millionth of a liter in some cases may cause an experiment to fail.

The laboratory activities have an accompanying results/questions sheet for recording the results of the experiment and answering questions related to the laboratory exercise. Looking at the questions prior to starting an exercise may help you focus on the key concepts of that exercise.

While this manual is designed to stand alone, the Photographic Atlas is a supplementary publication that provides additional information about the laboratory exercises along with photographs and diagrams that illustrate the procedures.

Good luck in your exploration of molecular biology. I have found it to be a most rewarding and fascinating area of biology. I hope you find it to be equally rewarding. If you have any comments on the laboratories or the exercises, please contact me c/o Morton Publishing Co.

Acknowledgments

I thank my wife, Audrey for her support during the writing of this manual. I thank Mr. Steve Plum who collaborated on the development of several of the exercises presented in this manual. I appreciated the constructive criticism of two reviewers, Dr. Stephanie Dellis and Dr. Mary Ellen Krause, whose comments substantially improved this book. I also thank my students in Molecular Biology, Molecular Genetics Laboratory, and Biotechnology for Teachers classes who have helped me refine and improve these exercises. Finally, I acknowledge the receipt of a NSF equipment grant (DUE ILI 9650215), which facilitated the development of many of these exercises.

Several companies donated materials that were used in the development of some of the exercises. This companies included: Ambion, Inc., Austin, TX, Epicentre Technologies, Madison, WI; FMC Bioproducts, Rockland, ME, and Promega Biotech, Madison, WI.

Contents

Preface iii

Acknowledgments iv

Safety vi

I Basic Techniques 1
Laboratory 1: Pipettor Use 3
Laboratory 2: Agarose Gel Electrophoresis 9

II DNA Analysis Techniques 15
Laboratory 3: RFLP Analysis 17
Laboratory 4: Complete RFLP Analysis; Southern Blotting of DNA 21
Laboratory 5: DNA Probe Preparation 27
Laboratory 6: DNA Probe Labeling and Testing 33
Laboratory 7: Pre-hybridization, Hybridization 39
Laboratory 8: Detecting λ DNA on the Membrane 45
Laboratory 9: Polymerase Chain Reaction (PCR) 51

III RNA Isolation and Analysis Techniques 57
Laboratory 10: mRNA Isolation 59
Laboratory 11: Reverse-Transcriptase PCR (RT-PCR) 65
Laboratory 12: Northern Blotting 69

IV DNA Manipulation and Cloning 77
Laboratory 13: Competent Cell Preparation and Transformation 79
Laboratory 14: Plasmid DNA Isolation 85
Laboratory 15: Linker-based Mutagenesis 93
Laboratory 16: pGEM®-*luc*, Linker DNA Ligation 95
Laboratory 17: Isolating Clones that Express Luciferase 99

V Advanced DNA Analysis 105
Laboratory 18: DNA Sequencing 107
Laboratory 19: Computer Analysis of DNA Sequence Information 115
Laboratory 20: Gel Retardation Assay 119

VI Protein Analysis 123
Laboratory 21: SDS-PAGE Analysis of Proteins 125
Laboratory 22: Western Blotting 131

Appendix: Preparation of Frequently Used Solutions 135

Index 137

Safety Considerations

Laboratory safety involves two things: 1) Knowing what hazards exist and 2) Acting appropriately to minimize those risks. In this section, general procedures are described that are designed to reduce the hazards associated with molecular biology research. In addition, for laboratory experiments that involve known hazards, specific information is given within the description of that exercise.

General Safety Procedures

1. Assume all chemicals and other reagents are hazardous and avoid any direct contact with them.

2. Always wear the appropriate protective equipment when dictated by the chemical and physical hazards of the experiment. For example, disposable gloves, laboratory coats, and eye protection should be worn when conducting some of these experiments.

3. Avoid hand to mouth operations in the laboratory, including eating, drinking, pencil-chewing, etc. This will reduce the chance that you will ingest toxic chemicals or other hazardous materials used in the laboratory.

4. Always use mechanical aids for pipetting. NEVER MOUTH PIPET! Nearly every laboratory exercise will involve some type of liquid transfer, so it is important you become proficient in the use of these devices.

5. The high voltages used in electrophoresis can be dangerous. Make sure you follow the guidelines for electrophoresis given by your instructor. In particular, make sure the power supply is turned off before opening or otherwise handling a gel apparatus.

6. You will use a ultraviolet transilluminator for visualizing DNA in gels. Make sure you are wearing UV-blocking goggles, gloves, and a lab coat or long sleeved shirt to protect yourself from the intense ultraviolet light given off by the transilluminator.

7. Always be sure that the microcentrifuge is balanced before you spin your samples. The rotor of the microcentrifuge needs to be balanced so that every tube has a "partner" on the opposite side of the rotor.

8. Wipe down your work area before leaving the laboratory and always wash your hands before leaving the laboratory.

9. Treat any microbial culture as potentially pathogenic. Discard any disposable material that contains bacteria in a biohazard waste container. Place test tubes and other reusable supplies that contain bacteria in an appropriate storage container for later autoclaving.

References

Biosafety in Microbiological and Biomedical Laboratories, 3rd Ed. 1993. Washington, D.C.: U.S. Govt. Printing Office.

Biosafety in the Laboratory: Prudent Practices for the Handling and Disposal of Infectious Material. 1989. Washington, D.C.: National Academy Press.

Laboratory Safety: Principles and Practices. 1995. Washington, D.C.: ASM Press.

Prudent Practices in the Laboratory: Handling and Disposal of Chemicals. 1995. Washington, D.C.: National Academy Press.

I

Basic Techniques

The laboratory exercises in this section introduce you to equipment and techniques that will be used throughout the course. In Laboratory 1, you will work through a basic exercise designed to teach you how to properly use a pipettor. In Laboratory 2, you will use agarose gel electrophoresis to determine the size of two DNA fragments.

Pipettors will be used in nearly every exercise in this laboratory manual; agarose gel electrophoresis will be used in many of the exercises. Developing an understanding of these methods and developing the dexterity to carry out these techniques early in the course will help ensure your success with the other exercises in the manual.

Pipettor Use

One of the most commonly used tools of molecular biology is the pipettor. This device is used to transfer very small volumes — 1/1000 to 1/1,000,000 of a liter. The purpose of this laboratory session is to teach you proper use of pipettors. A diagram of a pipettor is shown in Figure 1.1.

Some general comments about these devices:

1. They are expensive, precision instruments. Handle them carefully, being especially cautious to not drop them.

Tip ejector button

Thumb button for pipetting or expelling liquids

Thumbwheel for adjusting volume

Barrel — tips go on the extreme end

Figure 1.1   Diagram showing the main parts of a pipettor.

Photographic Atlas Reference Chapter 1

2. Never hold them or leave them sitting horizontally with liquid inside the tip — the liquid may run into the barrel of the pipettor, possibly gumming up the device, and/or contaminating your next sample.

3. Never adjust a pipettor outside its engineered range. For example, a pipettor with volume adjustments between 1 µl and 100 µl should not be adjusted to transfer a volume above or below those values. Doing so can jam the adjustment mechanism and destroy the pipettor.

4. You will be using these instruments in almost every laboratory session in this course. If you are not sure that you are using them properly, be sure to ask for help.

General Procedures for Use of Pipettors

There are many brands of pipetting devices. Therefore, your instructor will provide specific directions for the model you are using. As a general rule, though, pipettors have a dial or wheel for volume adjustments, a window that shows the volume setting, and a narrow end for holding a disposable pipet tip. The directions that follow are for the Gilson Pipetman®.

1. Adjust the volume on the pipettor by turning the thumbwheel (on newer models, the thumb button can also be used to adjust volume) until the appropriate volume shows in the window. With this pipettor, you get the most accurate measurements if you dial down to the volume. So, for example, if the pipettor is set for 5 µl and you need to transfer 10 µl, turn the dial so you get a reading of 12 or 15 µl in the window. Then dial back down to 10 µl.

2. Press a disposable yellow tip firmly onto the end of the pipettor.

3. Pipet up the appropriate volume of liquid by depressing the thumb button only until the point at which you feel resistance (not all the way to the handle), then put the yellow tip in the liquid. Finally, slowly raise the thumb button to its starting position; this causes liquid to flow into the yellow tip.

4. Remove the pipettor from the liquid and visually verify that an approximately correct volume has been pulled into the yellow tip. Dispense the liquid into the appropriate tube by pushing the thumb button all the way down to the handle. Avoid putting the pipet in a horizontal position during the liquid transfer.

5. If you are going to pipet a new solution, dispose of the yellow tip by depressing the small flange at the top of the handle.

6. Note that the disposable tips you will be using normally are sterile. Replace the cover on the box of tips immediately after you remove a single tip.

Option A: Use of Pipettor (See Figure 1.2).

Make sure a small plastic weigh boat is on the pan of the balance. Tare the balance.

Test the Calibration of the Pipettors

Pipet up 10 µl of distilled water with the p20. Pipet into the weigh boat. The balance should register 10 mg. Dry the weigh boat, place the weigh boat back on the balance, and re-tare if necessary.

Pipet up 100 µl of distilled water with the p200. Pipet into the weigh boat. The balance should register 100 mg. Dry the weigh boat, place the weigh boat back on the balance, and re-tare if necessary.

If your pipettors are off by more than 10% from the expected values, retest them. If they are still off after the second test, notify your instructor. The pipettor may need re-calibration, or there may be errors in your pipetting technique.

Measuring with the Pipettors

1. Pipet up 5 µl of a 40% sucrose solution with the p20. Pipet into the weigh boat. Adjust the volume, then pipet up 20 µl liquid with the p20, pipet into the same weigh boat that already contains 5 µl of the sucrose solution. Record the weight of the combined volume of 25 µl in the worksheet at the end

of this chapter. Dry the weigh boat, place the weigh boat back on the balance, and re-tare if necessary.

2. Now using the p200, pipet up 100 µl of 40% sucrose. Pipet into the weigh boat. Then adjust the volume, add 200 µl more liquid to the weigh boat. Record weight in the worksheet. Dry the weigh boat, place the weigh boat back on the balance, and re-tare if necessary.

Option B: Use of Pipettor

Precautions

1 M HCl is a STRONG ACID. Immediately rinse with water if you spill any on your skin; wear appropriate safety equipment when handling this solution as required by your institution.

1. Add 1 µl of 1 M HCl to a clean microcentrifuge tube, then add 50 µl of 0.1 M Tris to the tube. Mix well by pipetting up and down. Remove 20 µl of this mixture, and pipet to a small piece of pH paper. Record the pH in the worksheet at the end of this chapter.

2. Now add 2 µl of 1 M HCl and 50 µl of 0.1 M Tris to a new tube. Mix by pipetting up and down. Measure the pH as above. Record the pH in the worksheet.

3. Finally, add 50 µl of 1 M HCl and 1 µl of 0.1 M Tris to a new tube. Mix by pipetting up and down. Measure the pH as described above. Record the pH in the worksheet.

Discard the tubes as directed by your instructor.

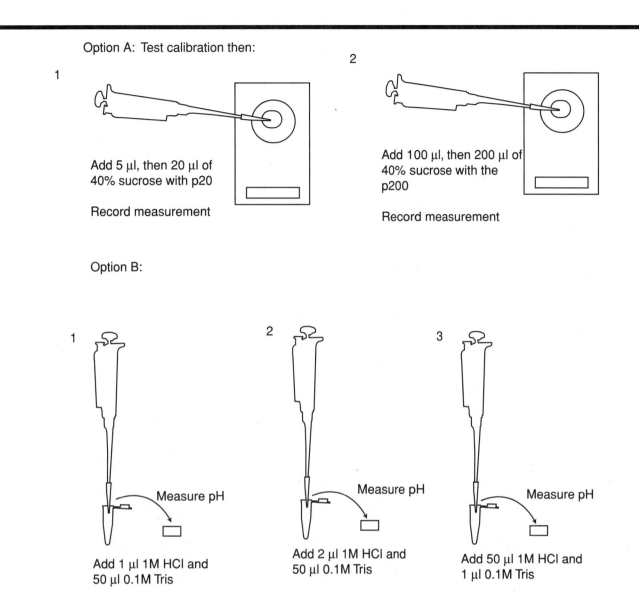

Option A: Test calibration then:

1

Add 5 µl, then 20 µl of 40% sucrose with p20

Record measurement

2

Add 100 µl, then 200 µl of 40% sucrose with the p200

Record measurement

Option B:

1

Measure pH

Add 1 µl 1M HCl and 50 µl 0.1M Tris

2

Measure pH

Add 2 µl 1M HCl and 50 µl 0.1M Tris

3

Measure pH

Add 50 µl 1M HCl and 1 µl 0.1M Tris

Figure 1.2   Outline of lab exercise 1. The two options for the lab are shown.

Pipettor Use

Name: _____   Date _____

Option A

1. Weight of the first volume   _____mg

2. Weight of the second volume   _____mg

Option B

1. Record the pH from the first measurement   _____

2. Record the pH from the second measurement   _____

3. Record the pH from the third measurement   _____

Agarose Gel Electrophoresis

Key Terms

Agarose: a highly purified polysaccharide, used for preparing gels that separate DNA (and RNA) molecules by size.

UV transilluminator: a high-intensity ultraviolet light source that is used for visualizing DNA stained with fluorescent dyes including ethidium bromide.

Ethidium bromide: a commonly used DNA stain. Ethidium bromide binds to DNA. It absorbs ultraviolet light, and gives off orange visible light that can be easily detected.

Tris: (Tris[hydroxymethyl]aminomethane) a commonly used buffer in molecular biology. Tris has a pK_a of 8.1 and a useful buffering range between ~ pH 7 - 9.

EDTA: (Ethylenediaminetetraacetate) a metal chelator that is frequently added to solutions containing DNA. EDTA tightly binds metal ions like Mg++ that are required cofactors for the activity of DNases (enzymes that degrade DNA).

TBE buffer: An electrophoresis buffer containing Tris, boric acid, and EDTA.

TE buffer: A DNA resuspension buffer containing low concentrations of Tris and EDTA.

Loading (or tracking) buffer: A solution added to DNA before it is loaded into an agarose gel. This buffer facilitates loading DNA by making the DNA solution more dense than the electrophoresis buffer. The loading buffer also contains one or more dyes that permit one to gauge the progress of electrophoresis.

Materials

- Electrophoresis apparatus — power supply, gel box, gel tray, comb
- Agarose
- 0.5X TBE buffer
- p20 pipettor, tips for the pipettor
- DNA samples (e.g., λ *Hind*III digest and an unknown DNA, each containing 0.5µg to 1 µg/12 µl of DNA and diluted with TE and loading buffer)
- Ethidium bromide solution (5 mg/ml)
- UV transilluminator
- Instant camera with type 667 film
- Microwave

Photographic Atlas Reference
Chapter 2

Agarose gels are widely used for separating biomolecules, particularly DNA. The agarose gel contains innumerable pores. When DNAs are driven through the gel by an electrical current, small DNA fragments migrate more quickly through the pores compared to large DNA fragments. This differential migration allows for the separation of differently-sized DNA molecules.

In this laboratory exercise, you will get additional practice with your pipettor as well as gain experience performing agarose gel electrophoresis. Specifically, you will be using the agarose gel to determine the sizes of different DNA fragments.

Precautions

- Handle heated agarose solutions with insulated gloves to prevent burns. Agarose, particularly when heated in a microwave, can become superheated and boil violently when handled.

- Always turn off the power and disconnect the leads before removing or inserting an agarose gel into the gel box. The power supplies used in electrophoresis can deliver a potentially lethal shock.

- Handle ethidium bromide stain and the stained agarose gels with gloves. DNA stains including ethidium bromide are potential or known mutagens and potential carcinogens.

- Wear gloves, long sleeves, and UV-blocking goggles when using the UV transilluminator. This will help prevent a severe "sunburn" that can result from even short exposures to the UV light from the transilluminator.

- The concentrated HCl required for the preparation of TE buffer is potentially hazardous. **Wear gloves, goggles, and a lab coat and add the HCl to the solution in a chemical fume hood.**

Procedures

In this laboratory, you will be preparing an agarose gel and running a sample of DNA in the gel. The directions below are for minigel apparatus (20–25 ml gel volume), e.g., OWL B1A apparatus. For other electrophoresis apparati, follow your instructor's guidelines.

Preparation of Solutions

Consult the appendix for the preparation of the solutions required for this exercise.

1. Set up your gel mold so the gel will be ready to pour. For the Owl apparatus, turn the gel mold 90° in the gel box, making sure the gaskets form a continuous seal. Place an 8 or 10-well comb in the appropriate slot of the gel mold. Other gel molds either are sealed with some type of flexible plug or with waterproof tape. One gel will be needed per two or three students or student groups. The table where the gel is to be poured must be level, otherwise one part of the gel will be thicker than the other.

2. Make a 0.8% agarose gel. Add 25 ml of 0.5X TBE buffer to a loosely-capped, microwave-safe bottle. (You must use TBE buffer, not water, or the gel will not run properly.) Add 0.2 g agarose. Swirl to mix. Place in the microwave with the bottle loosely-capped. Heat at 50% power for 1–2 min. *Caution, to prevent boil over and to ensure the agarose does not become superheated, stop the microwave every 10–15 seconds and, wearing heat-resistant gloves, swirl the container.* The solution is ready to use if no transparent "lenses" are visible when the agarose is swirled and held up to the light. Wearing gloves, add 2.5 µl of a 5 mg/ml ethidium bromide solution to the melted agarose. Swirl to mix. Pour slowly into a gel mold. If bubbles are present on the surface of the gel, move them off to the side with the gel comb. Allow the gel to sit for at least 30 min., until it is solidified. Solid gels have a translucent appearance; liquid agarose is transparent.

3. When the gel has solidified, pour 0.5 X TBE buffer over the gel. Cover the gel with buffer to a depth of about 2 to 3 mm. Then carefully remove the comb.

Loading/Running/Analyzing Gels

1. Each person should load two lanes of a gel. Load 12 µl of the unknown DNA solution to one lane, 12 µl of the known DNA to the other lane. (The DNA solution should contain 0.5 to 1 µg of DNA and the appropriate volume of TE and loading dye to equal 12 µl. For a typical DNA stock solution with a concentration of 0.5 µg of DNA/µl, you will need 1–2 µl of the DNA stock solution, 2 µl of 6X loading dye, and 8–9 µl of TE buffer.)

The wells in the gel are like a trough. Insert your pipet tip just into the trough and pipet out the liquid slowly (Figure 2.1). If you put your pipet tip in too deeply, you can puncture the bottom of the well. Be sure to keep the pipet plunger depressed until your tip is completely out of the electrophoresis buffer. Otherwise you may draw your sample back into the pipet tip. It is often useful to rest one hand on the gel box to support your other hand to prevent the pipettor from wiggling too much.

2. When everyone is done loading the gel, attach the leads from the electrophoresis chamber to the power supply. Be sure the positive pole (anode, usually a red lead) is at the bottom of the gel, because negatively charged DNA migrates toward the positive pole. Turn on the power. The proper voltage to apply will vary with the sizes of DNA you are trying to resolve. For this exercise, the voltage should be set at about 5V/cm (cm referring

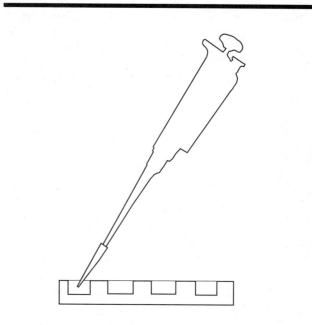

Proper placement of pipet tip for gel loading

Figure 2.1 Detail of agarose gel loading technique, showing side view of a gel.

to the distance between electrodes). *For an OWL B1A gel apparatus (16 cm between electrodes), electrophoresis should be performed at about 90V.* Check the gel after about 5 minutes to make sure the DNA is migrating in the correct direction. The gel should be electrophoresed until the dark blue dye has traveled one-half to two-thirds the length of the gel. This will take 45 minutes or more, depending on your apparatus.

3. Since the stain is in the gel, the DNA will be stained as it is electrophoresed, and the gel can be viewed immediately after electrophoresis by placing it on an UV transilluminator. (Always wear gloves when handling the gel, and follow the appropriate safety precautions when using the UV transilluminator.) To measure the migration of the DNA fragments accurately, place a ruler next to the gel and photograph the gel and the ruler with a digital camera or a Polaroid™ instant camera with Type 667 film. Generally an f-stop of f5.6 to 8, with the shutter speed set for 1 second, works well for most instant cameras. Be sure your camera has the appropriate filter.

   If no camera is available, measure the distance the bands have migrated by placing a ruler on the transilluminator next to the gel and recording the measurements. In either case, measure the DNA consistently, typically measuring the distance from the middle of the well to the middle of each band.

4. Graph the size of the DNA vs. the distance migrated on the worksheet. Note that on semi-log paper there is nearly a linear relationship between the distance the DNA migrates and the log of the size of double-stranded DNA (in base pairs). The size of the DNA fragments from the λ *Hind*III digest are: 23,130, 9,416, 6557, 4361, 2322, 2027, 564 (there is also a 125 bp band which may be visible under optimal conditions).

Tips

1. The longer the DNA is allowed to migrate in the gel, the easier it will be to accurately size the DNA fragments, because the DNAs are separated by a greater distance.

2. Agarose needs to be completely melted (no lenses). If not, the un-melted agarose will conduct the electrical current differently from the surrounding matrix, resulting in V-shaped DNA bands.

3. For heating agarose, use 50% power for 30–45 sec. with the lid of the container loosened. Wear oven mitts — stir frequently during microwaving — stop if the solution starts to boil; mix, microwave, stop, mix, microwave, stop, etc.

4. Gels must be completely solidified before the comb is removed, otherwise the wells will be misshapen and the DNA bands will be distorted.

5. Be sure leads are properly placed — DNA being negatively charged, will migrate toward + pole. Check the gel after about 5 min. to make sure the DNA is migrating in the right direction.

6. Don't overload the gel, and make sure the gel is made with buffer (not water!).

7. Agarose gel electrophoresis is used in many experiments in this laboratory manual. Always follow the safety guidelines for electrophoresis described in this exercise.

References

The Sourcebook. 1998. Your complete guide for DNA separation and analysis. Rockland, ME: FMC Bioproducts

Sambrook, J., E. Fritsch, and T. Maniatis. 1989. Molecular cloning. A laboratory manual. 2nd Ed. Cold Spring Harbor, NY: Cold Spring Harbor Laboratory Press: Chap. 6.

Agarose Gel Electrophoresis

Laboratory 2

Name:_____ Date _____

Results

Plot the migration of the DNAs on the semi-log paper on the following page. Plot the distance migrated (in mm) along the X-axis. Plot the size of DNA (in bp) along the logarithmic Y-axis.

Questions

1. Based on your graph, how large is the first "unknown" DNA band identified by your instructor?

2. Based on your graph, how large is the second "unknown" DNA fragment identified by your instructor?

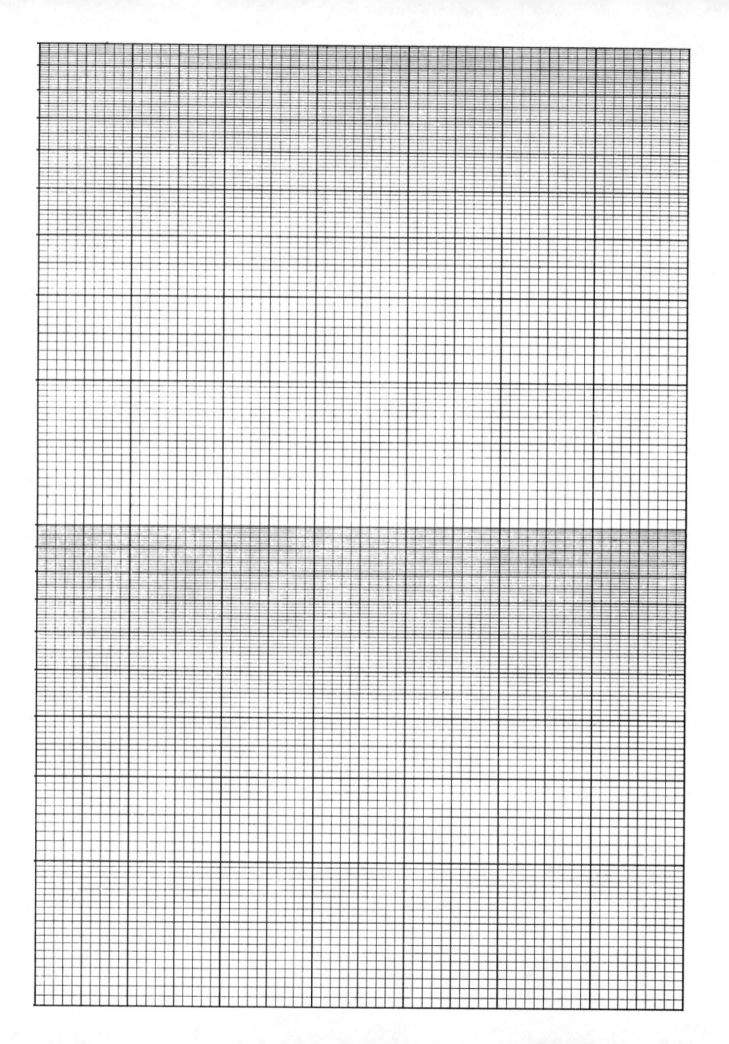

DNA Analysis Techniques

The exercises in this section of the laboratory manual are designed to teach you several basic techniques that are routinely used for the analysis of DNA. Starting with the technique of agarose gel electrophoresis that you learned in Laboratory 2, you will now use gel electrophoresis to identify differences in the size of restriction enzyme fragments from the digests of different bacteriophage genomes (Laboratory 3 and 4). This is Restriction Fragment Length Polymorphism (RFLP) analysis, and the technique is used to study differences between the genomes of different organisms, and to identify DNA fragments that have been cloned into plasmids or other vectors.

The bacteriophage DNA (λ and λgt10) that was separated in the gel will be analyzed, then transferred to a nylon membrane (Laboratory 4). This technique, Southern blotting, preserves a copy of the DNA pattern originally present in the gel. λgt10 was genetically engineered for cloning cDNAs. During the course of this genetic engineering, a piece of DNA was deleted. Laboratories 5 through 8 are designed to show you how one can identify which section of DNA has actually been deleted from the phage genome. In Laboratory Exercise 5, you will isolate a fragment of DNA that will be used as a probe. You will label that DNA with a molecule called digoxigenin during Laboratory Exercise 6. The blotted DNA will then be incubated with the labeled DNA probe (Laboratory 7), and bound probe will be identified (Laboratory 8) to determine if there are detectable differences in the DNA sequences of the two phage DNAs that have been immobilized on the membrane.

In the final laboratory exercise in this section, you will do a Polymerase Chain Reaction (PCR) analysis of DNA from your epithelial cells. PCR is a very powerful technique for rapidly amplifying sections of DNA for further analysis. In Laboratory 9 you will amplify DNA by PCR, then compare the size of the fragments amplified by PCR with the PCR fragments from your classmates. This is a more rapid method of RFLP analysis, sometimes called AFLP (Amplified Fragment Length Polymorphism) analysis.

PCR is also used for many other purposes including the amplification of DNA for cloning, for DNA sequencing, and for introducing mutations at specific sites in a DNA sequence.

RFLP Analysis

Laboratory

3

Key Terms

Restriction enzymes: Bacterial proteins that cut double-stranded DNA at specific sequences. You will be using the restriction enzyme *Eco*RI in this laboratory exercise, which recognizes the DNA sequence 5'-GAATTC-3'.

RFLP analysis: Restriction Fragment Length Polymorphism analysis. This technique involves comparing DNA molecules based on the size of the fragments produced after digestion with a restriction enzyme.

Bacteriophage λ: Bacteriophages are viruses that attack prokaryotes. Bacteriophage λ is a phage that specifically attacks certain strains of *Escherichia coli.*

Genome: The entire DNA complement of an organism.

Recombinant DNA: Generating a novel DNA molecule by joining two DNAs, often from unrelated organisms.

Microcentrifuge tubes: Small plastic tubes used spinning down materials in a small centrifuge. Microcentrifuge tubes typically hold either about 0.5 or 1.5 ml of liquid. In addition to their use in centrifugation, microcentrifuge tubes are also often used as reaction vessels for small volumes of liquids.

Materials

- Phage A — λgt10 DNA — 17 µl (0.2 to 0.5 µg of λgt10 DNA in TE)
- Phage B — λ DNA — 17 µl (0.5 to 1 µg of λ DNA in TE)
- Phage U — unknown — 17 µl (λ or λgt10 DNA in TE)
- Three, 0.5 ml or 1.7 ml microcentrifuge tubes
- *Eco*RI restriction enzyme, restriction enzyme buffer
- p20 pipettors, pipet tips
- Agarose
- Electrophoresis apparatus
- 0.5X TBE buffer
- Ethidium bromide solution

Photographic Atlas Reference
Chapter 3

Restriction enzymes are widely used in molecular biology for both analyzing DNA and creating recombinant DNA molecules. In this laboratory, you will have an opportunity to see how these enzymes are used for analyzing DNA. In this exercise, you will analyze restriction enzyme DNA digests of two slightly different bacteriophage genomes.

Precautions

Follow the directions given in laboratory 2 for proper preparation and handling of ethidium bromide-stained agarose gels.

Procedure

Restriction Enzyme Digestion

1. Add 2 µl of *Eco*RI digestion buffer to each of the three, 0.5 ml microcentrifuge tubes (not PCR tubes).

2. Add 17 µl of phage A DNA to one 0.5 ml microcentrifuge tube. Label the tube with your initials and "A".

3. Add 17 of µl phage B DNA to one 0.5 ml microcentrifuge tube. Label the tube with your initials and "B".

4. Add 17 µl of phage "U" DNA to the third microcentrifuge tube. Put your initials and "U" on tube.

5. Add 1 µl of *Eco*RI restriction enzyme to each tube. Put the tubes into a rack. Incubate at 37°C for 1 to 1.5 hours or overnight. Then store the samples at −20°C until they are ready to be run on a gel.

6. λgt10 has a total length of 43,340 bp and has an *Eco*RI restriction enzyme site located at base pair 32,710. λ has a total length of 48,500 bp, with

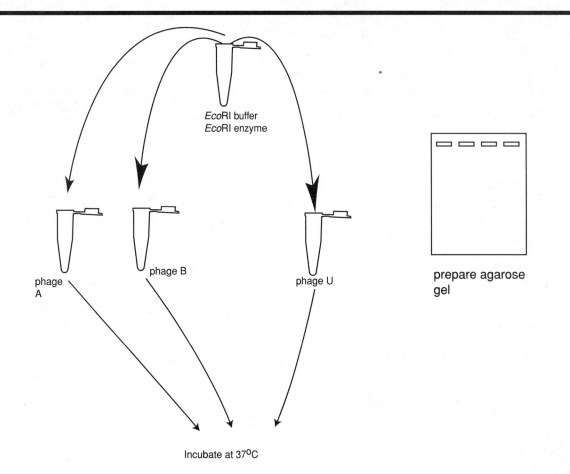

EcoRI buffer
EcoRI enzyme

phage
A

phage B

phage U

prepare agarose
gel

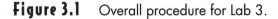

Incubate at 37ºC

Figure 3.1 Overall procedure for Lab 3.

EcoRI sites at bp 21,230; 26,100; 31,750; 39,170; and 44,970. Sketch what you expect in terms of the sizes of EcoRI-digested λgt10 and λ DNAs in the worksheet at the end of the chapter.

Agarose Gel Preparation

(See appendix for preparation of solutions, if required).

While the digest is incubating, prepare a 0.6% agarose gel.

1. Pour 25 ml 0.5X TBE into a microwave-safe, loosely capped bottle.

2. Add 0.15 g of agarose. Put the cap back on loosely.

3. Heat in microwave at 50% power for about 2 min. with frequent swirling. Make sure agarose is completely dissolved. Make sure the agarose is not superheated before handling the bottle; superheated agarose can boil violently and burn you.

Add 2.5 µl of a 5 mg/ml ethidium bromide solution to the melted agarose, swirl to mix.

5. Pour into gel mold. Make sure the appropriate combs are in place before pouring the gels.

6. Allow the gels to solidify.

If the gel is not going to be used immediately, it can be tightly wrapped in plastic food wrap and stored in the refrigerator several days.

Reference

Guilfoile, P. and S. Plum. 1998. An authentic RFLP lab for High School or College Biology Students. American Biology Teacher 60(6): 448- 452.

RFLP Analysis

Name:_____ **Date** _____

Results

Questions

1. Sketch the expected sizes of the λ and λgt10 DNAs in the box below. Show the fragment sizes (in bp) along the left side of the gel.

2. Was your unknown DNA λ or λgt10? How did you know?

Key Terms

Southern blotting: (also called Southern transfer) a method for transferring DNA from an agarose gel to a membrane. Once the DNA is bound to the membrane it can be used in experiments designed to test the similarity of the DNA on the membrane with other, labeled DNAs (Labs 6–8).

20X SSC: a buffered, high-salt solution used to wash membranes after blotting and during DNA hybridization procedures.

cohesive ends: Complementary, single-stranded 12 base-pair ends of λ DNA. When λ DNA is digested, these complementary ends can rejoin, potentially complicating analysis of the banding pattern of the DNA.

In this laboratory exercise, you will load your digested samples from laboratory 3, analyze the restriction enzyme digestion patterns, then transfer the DNA to a nylon membrane for further analysis by

Photographic Atlas Reference Chapter 4

DNA hybridization (Southern blotting and hybridization — Labs 7 and 8).

Precautions

- Be sure to follow the directions given in laboratory 2 for proper procedures to use when handling ethidium bromide-stained agarose gels and when performing electrophoresis.

- The transfer buffer contains a strong base (NaOH). Wear the appropriate safety equipment and follow the appropriate safety precautions, as directed by your instructor, when handling the transfer buffer.

- Be sure the microcentrifuge is balanced before centrifuging your samples.

Materials

For Gel Electrophoresis:

- Previously digested DNA samples
- Marker DNA with band sizes between ~ 2000 and 20,000 bp (e.g., λ *Hin*dIII digest)
- p20 pipettor, yellow tips
- loading buffer (typically supplied with DNA markers, or prepared as described in the Appendix)
- agarose gel (preparation described in the previous exercise)
- 0.5X TBE buffer (prepared as directed in the Appendix)
- Electrophoresis apparatus

For DNA Transfer:

- Glass or plastic dish
- Glass plate
- Plastic food wrap
- Nylon membrane cut to gel size
- Transfer solution
- Gel with DNA
- 2 pieces of Whatman 3MM paper cut to gel size
- 1 Whatman 3MM paper "wick"
- ~50 pre-cut paper towels cut to gel size
- small weight (e.g., 200 ml plastic bottle with water)
- 20X SSC (prepared as described in the appendix)

Preparation of Solution
(See precautions, above)

Procedures (See Figure 4.1)

Loading the Gels with Your DNA Samples

Optional: Heat the digested DNAs for 5 min. at 55°C. This will disrupt bonds between the cohesive ends of λ and λgt10 DNAs and result in bands with more consistent intensities. Place on ice for 1–2 min. after heating.

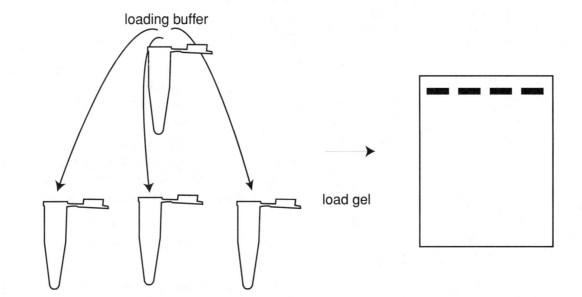

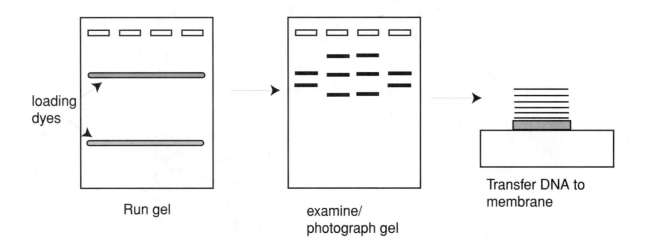

Figure 4.1   Overall procedure for Lab 4.

1. Spin the samples briefly in the microcentrifuge to pull down condensation. *(Be sure to balance the rotor before centrifugation.)*

2. Add 3 µl of 6X loading buffer to your samples.

3. Load your entire A, B, and U samples into the gel, in that order. One marker DNA (λ DNA digested with *Hind*III) should be loaded on each gel as well.

4. The gel will be run for about 1 to 1.5 hours (3V/cm). After the run, place the gel on the UV transilluminator and photograph it. Leave the gel in its tray, if possible, since this gel is very fragile. While the gel is running, set up the DNA transfer apparatus described below.

Transfer of DNA to Nylon Membranes

1. Prepare materials for blotting DNA to a nylon membrane. One to four gels can be blotted with a single transfer set-up.

a. Cut a strip of Whatman paper as wide as the gel is long. The paper needs to be long enough to span the glass plate and touch the bottom of the dish containing transfer buffer on each side of the glass plate (Figure 4.2).

b. While wearing gloves, cut one piece of nylon membrane to the size of the gel. Only touch the edge of the membrane, preferably with a blunt forceps while wearing powder-free gloves, and handle the membrane as little as possible.

c. While wearing gloves, cut 2 pieces of Whatman 3 MM filter paper to the same size as the gel.

d. Cut 50 paper towels to the size of the gel. Cut 4 pieces of plastic food wrap about 3 cm wide. Each piece should be about 3 cm longer than the gel in its longest dimension.

e. Add transfer buffer to the dish and saturate the long strip of Whatman 3MM paper. (This strip of paper will act as a wick to transfer liquid from the dish through the gel and into the

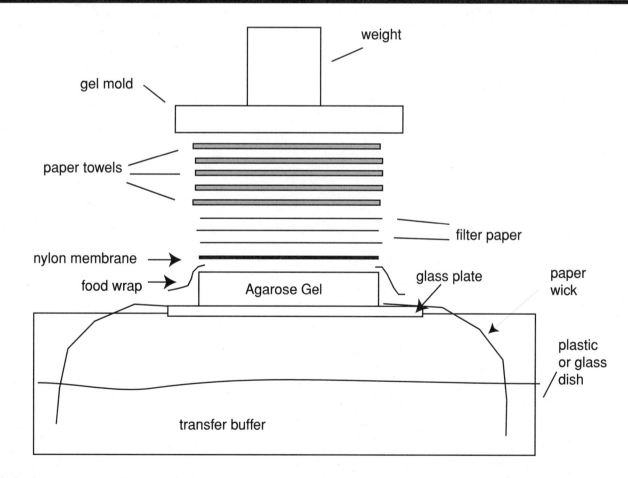

Figure 4.2   Southern transfer.

membrane and paper stack. This wicking action will cause the DNA to move out of the gel and on to the membrane. The NaOH in the transfer buffer will denature the DNA, causing it to become single stranded. The DNA on the membrane must be single-stranded, or it will not bind to the DNA probe in Laboratory Exercises 7 and 8.)

2. After the agarose gel has been run and photographed, carefully slide the gel out of the gel mold on to the Whatman paper wick (the gel can easily break if you handle it roughly). The wick should be completely saturated with buffer. After putting the gel down on the blotting paper wick, roll a clean pipet over the gel to remove air bubbles trapped between the gel and the paper. (Air bubbles will prevent the transfer of DNA to the membrane directly above the air bubble.) Lay a strip of plastic food wrap along each side of the gel. Cover the gel edge for about 2–4 mm. The food wrap should be a barrier around the gel that prevents the nylon membrane or the Whatman paper from contacting the wick. Lay the nylon membrane smoothly and carefully on top of the gel; roll out the air bubbles with a pipet. Place the pieces of Whatman paper on top of the membrane, one at a time, rolling out the bubbles after each piece of paper. Next, place the 50 pieces of paper towels on top of the gel, followed by the gel mold and a small weight (e.g., a 200 ml plastic bottle filled with water) (Figure 4.2).

3. The transfer will be complete by the next morning. One member of each "gel group" needs to come in the next day to remove the membrane. Write on the bottom, back side of the membrane in pencil, so you can identify the DNA side and the top of the blot. Also, make sure the wells are marked with pencil. Then rinse the blot for a few seconds in 20 X SSC, UV-fix the membrane by placing on UV-transilluminator (DNA side closest to the UV light) for 3 minutes, then put the membrane on a piece of plastic food wrap and allow it to dry in a drawer. Dismantle the blotting apparatus. The blot will be used in Exercise 7.

Reference

Sambrook, J., E. Fritsch, and T. Maniatis. 1989. Molecular cloning. A laboratory manual. 2nd Ed. Cold Spring Harbor, NY: Cold Spring Harbor Laboratory Press: Chaps. 2, 9.

Complete RFLP Analysis

Name: _____   Date _____

Results

Questions

1. In the optional heating step, why would heating the DNA to 55°C disrupt bonding of the cohesive ends, but not disrupt H-bonding of the rest of the DNA molecules in the DNA solution?

2. If you had used plain water rather than NaOH as the transfer buffer, you would not have detected a signal on the membrane after hybridization. Why not?

3. When large DNA fragments are being transferred, sometimes the gel is first treated with a dilute hydrochloric acid solution, then sodium hydroxide. This cleaves the sugar-phosphate backbone. Explain why this might be done.

Key Term

DNA probe: A fragment of DNA that is used for detecting specific DNA sequences, similar or identical to the sequence of the DNA probe. A probe needs to be labeled by some means, a procedure that is described in the next laboratory exercise.

In this laboratory exercise, you will be purifying DNA fragments that will be labeled in the next laboratory exercise.

Precautions

Follow the standard precautions for handling gels stained with ethidium bromide and for using the UV transilluminator as described in Laboratory 2.

Materials

- 0.5 X TBE buffer (prepared in laboratory 2; directions in Appendix)
- Electrophoresis apparatus
- Agarose
- Ethidium bromide solution
- GenElute™ Agarose Spin column (Supelco Inc. Bellefonte, PA, also available from Sigma Chemical Co.)
- TE buffer (prepared in laboratory 2; directions in Appendix)
- 7.5 M Ammonium acetate
- Isopropanol
- p20, p200 pipettors, yellow tips
- DNA size standard (e.g., λ digested with *Bst*EII is recommended)
- λ digested with *Hind*III

Photographic Atlas Reference
Chapter 4

Preparation of Solutions

7.5 M Ammonium acetate

To make 50 ml, add 28.9 g of ammonium acetate to 30 ml of distilled water. Mix to dissolve. Adjust volume to 50 ml by adding distilled water. Some authors recommend sterilization by filtration, but I have generally found this to be unnecessary.

Procedures

Isolation of DNA fragments

1. Prepare and run a 0.8 % agarose gel in 0.5X TBE buffer containing ethidium bromide as described in the previous lab exercises. For a 25 ml gel, this will require 0.2 g of agarose, 2.5 μl of a 5 mg/ml ethidium bromide solution, 25 ml of 0.5X TBE. Run the gel at 5V/cm (90V for a gel box with a 16 cm spacing between electrodes). Each group should run one sample consisting of 2 μg of λ DNA digested with *Hind*III, the proper volume of loading dye and TE buffer, in a total volume of 10 to 20 μl. Each gel should also have one marker DNA (λ *Bst*EII recommended).

   While the gel is running, prepare at least 1, 1% agarose-TBE-ethidium bromide gel; one lane is required for each λ*Hind*III digest run on the 0.8% gels. Use the narrowest comb available to form the wells on these gels. These gels will be used to verify the isolation of the proper DNA bands. If these gels will not be used during the same lab period, they should be wrapped tightly in food wrap and stored in the dark, in the refrigerator.

2. Prepare a spin column when the bromophenol blue dye has migrated about 1/2-way through the gel,

indicating the DNA has been separated. Add 100 μl of TE to the spin column. Place spin column in an open microcentrifuge tube. *Be sure the centrifuge is balanced with an equal number of tubes on either side of the rotor before the centrifuge is started. Use an extra tube to balance the centrifuge, if necessary.* Spin for 5 seconds at 14,000 rpm. *Pour off TE from microcentrifuge tube.* Set the spin column and the tube aside.

3. Visualize the DNA using the UV transilluminator. Work as quickly as possible to minimize UV damage to the DNA. Cut out the 2322 and 2027 bp bands with a razor blade. See Figure 5.1 for a schematic illustrating which bands to remove. The two, 2000 bp λ*Hin*dIII bands should be close together, relatively far down in the gel. One band from the λ*Bst*EII marker should migrate at almost exactly the same position as the 2322 bp band. Cut the bands in such a way as to minimize the amount of agarose in each band. You may find it useful to cut the gel just above and below the bands of interest, turn off the UV light, then cut the gel perpendicular to the initial two cuts. You should now have cut a rectangle out of the gel. Scoop up the gel piece with a corner of the razor blade and place the agarose slice containing the DNA into the Spin column. Turn the UV light back on to verify that the correct bands were removed. Spin the agarose gel slice for 10 minutes at 14,000 rpm in the microcentrifuge.

4. Discard the spin column, *keep the flow-through* (the liquid in the bottom of the microcentrifuge tube, which should contain the DNA). Add 1/2 volume of 7.5 M ammonium acetate and 1 volume of isopropanol to the flow-through. For example, if your flow-through volume is 100 μl, add 50 μl of 7.5 M ammonium acetate and 150 μl isopropanol. Mix by inverting and re-inverting the tube 10 to 15 times. Place on ice 10 min., spin 10 min. at 14,000 rpm in the microcentrifuge. Pour off the liquid after the tube has been centrifuged. Re-centrifuge for a few seconds, and remove all the liquid from the tube with a p200 pipettor, without disturbing the DNA pellet. Resuspend the DNA in 10 μl TE.

(If your lab period is only two hours long, it is likely the next step will have to be performed during the next lab period. In that case, store the purified DNA in a –20°C freezer until you need it.)

5. Mix 2.5 μl of DNA solution (prepared in Step 4, above), 6.0 μl TE, 1.5 μl of 6X loading buffer and run in one lane on the gel. λ*Hin*dIII should be run in one lane as well, so you can compare the size of the isolated bands with the expected sizes. If the isolation was successful, store the DNA at –20°C until the following laboratory session. If no DNA is visible in the lanes from the band isolation steps, repeat the procedure, this time using 4 μg of λ*Hin*dIII DNA per preparation. Make sure you follow the directions carefully, paying particular attention that the 7.5 M ammonium acetate was prepared correctly, and that you added the correct volume of ammonium acetate and isopropanol to the flow-through.

Tips

It is critical to cut as much agarose as possible away from the DNA. It is also important to minimize the exposure of the DNA to UV light.

As a general practice, when precipitating DNA, always place the hinge of the microcentrifuge tube up. That way, you will always know that any DNA pellet should be located on at the bottom of the tube, on the same side as the hinge, and you won't accidentally dislodge it when you pour off or pipet off the supernatant.

Reference

Protocol for Fast Recovery of DNA from Agarose Gel, Using GenElute™ Agarose Spin Columns. 1995. Supelco: Bellefonte, PA.

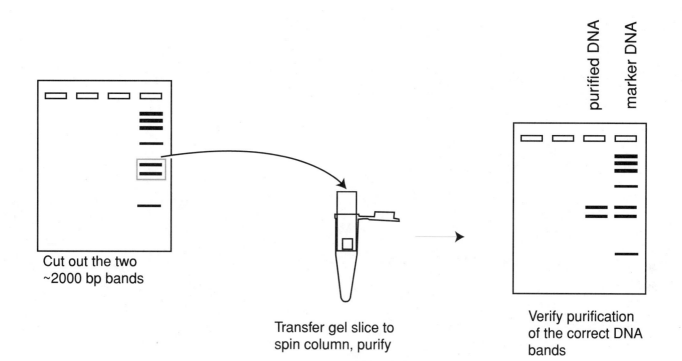

Cut out the two
~2000 bp bands

Transfer gel slice to
spin column, purify

Verify purification
of the correct DNA
bands

Figure 5.1 Overall procedure for Lab 5.

DNA Probe Preparation

Name:_____   **Date** _____

Results

Questions

1. What would be the expected result if you omitted the isopropanol from the reaction, after you used the spin column to purify the DNA?

2. Why was it important to check the DNA on another agarose gel after the purification process?

DNA Probe Labeling & Testing

Laboratory 6

Key Terms

probe: A labeled DNA molecule used to identify similar or identical DNA sequences on a membrane.

labeling: A method for introducing detectable chemicals into DNA. In this laboratory exercise, the DNA will be labeled with digoxigenin.

digoxigenin: A steroid compound isolated from plants of the genus *Digitalis*. DNA labeled with digoxigenin can be detected using an antibody that binds only to digoxigenin.

Klenow enzyme: A fragment of *E. coli* DNA polymerase I. This enzyme synthesizes DNA from nucleotide precursors. If the nucleotides contain digoxigenin, the DNA made from those nucleotides will be labeled.

hexanucleotide mix: A solution containing 6-nucleotide, single-stranded DNAs of random sequence. DNA synthesis by Klenow enzyme requires a primer (a short single-stranded DNA molecule). The hexanucleotide mix contains primers for this reaction, some of which will bind to any DNA and provide a starting point for DNA synthesis.

Materials

- DNA solution for labeling (1/2 of the DNA isolated in the previous lab exercise)
- Microcentrifuge tube with lid-lock (the lid lock prevents the tube from popping open during boiling)
- p20 pipettor, tips for pipettor
- Sterile, distilled or deionized water
- Boiling water bath
- 1 µl 0.5 M EDTA solution

The following reagents from Boehringer Mannheim (Indianapolis, IN)

- 2 µl Hexanucleotide mixture (Cat. #1277 081)
- 2 µl dNTP labeling mixture (Cat. #1277 065)
- 1 µl Klenow enzyme (Cat #1008 404)
- Digoxigenin-labeled DNA (a control for testing your probe; Cat #1 585 738, or DIG control teststrips Cat #1 669 966)

Photographic Atlas Reference
Chapter 4

Originally, DNA probes were labeled with radioactive nucleotides. In recent years, techniques have been developed for non-radioactive labeling of DNA using digoxigenin, biotin, or other compounds. In this laboratory exercise, the DNA fragments you isolated in Lab 5 will be labeled with digoxigenin. The probe will then be tested to verify that it is properly labeled. This probe will be used to detect DNA on the membrane in labs 7 and 8.

Precautions

Make sure that you do not touch or otherwise contact laboratory chemicals.

Procedure

(Adapted from "The DIG System User's Guide for Filter Hybridization.")

1. In a microcentrifuge tube, mix together
   - 5 µl of the DNA to be labeled (from the previous laboratory exercise)
   - 10 µl of distilled or deionized water.

   Boil the microcentrifuge tube (with a lid-lock) for 5 min. Place immediately on ice. (The boiling denatures the DNA (makes it single-stranded). If the DNA was not denatured, the hexanucleotide primers (added in the next step) could not bind to the DNA, and no labeled DNA would be synthesized.)

2. With the tube still on ice, add
   - 2 µl hexanucleotide mixture
   - 2 µl dNTP labeling mixture.

   Centrifuge the tube for 5 sec. in a microcentrifuge to spin down the condensation.

3. *Add 1 µl Klenow enzyme*, then incubate the microcentrifuge tube in a 37°C water bath or air

33

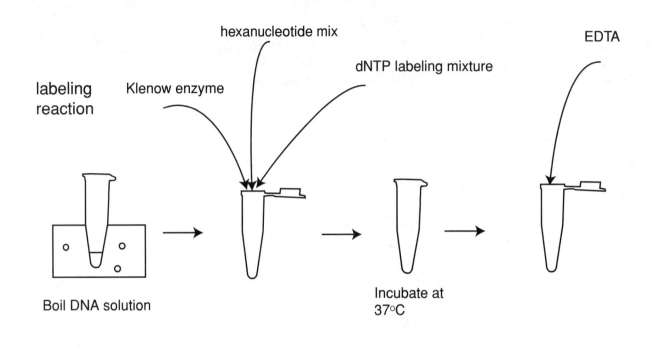

labeling reaction Klenow enzyme hexanucleotide mix dNTP labeling mixture EDTA

Boil DNA solution Incubate at 37°C

Figure 6.1 Procedure for the probe labeling reaction.

incubator for 1 hour. (Labeled DNA is being made during this incubation.)

4. At the end of the hour, add 1 μl EDTA solution. Store the probe in −20°C freezer until needed. (The EDTA binds magnesium ions; the Klenow enzyme requires magnesium ions for activity. Therefore, adding EDTA stops the Klenow-mediated labeling reaction.). The labeling reaction is depicted in Figure 6.1.

Tips

In Step 3, you can continue the incubation with Klenow enzyme for 20–24 hours before adding the EDTA; this can increase the yield of labeled DNA 3–4 fold.

For this laboratory exercise, the probe DNA does not need to be purified. In cases where the amount of detectable DNA is very low (e.g. identifying a single-copy gene from mammalian genomic DNA), it is useful to purify the reaction mixture on a column or by ethanol precipitation, in order to reduce background caused by unincorporated nucleotides. The simplest generally applicable method is to precipitate the

Materials

- Small piece of nylon membrane (Boehringer-Mannheim or Amersham Hybond N)
- Labeled probe from exercise above
- TE buffer
- Skim milk
- Promega Western Blue® or color substrate solution (appendix)
- Labeled control DNA (Boehringer Mannheim Cat. #1585738)
- Anti-digoxigenin antibody conjugated to alkaline phosphatase (Boehringer Mannheim Cat. #1093274)
- Maleic Acid buffer
- Washing buffer
- Blocking solution
- Detection buffer

(The preparation of the buffer solutions is described in the Appendix)

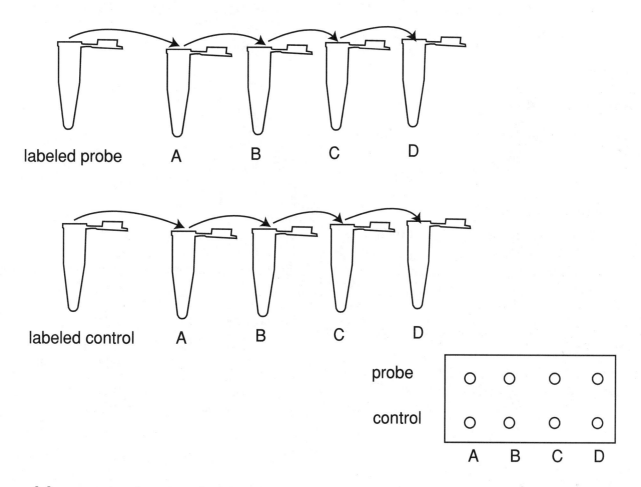

Figure 6.2 Procedure for testing the labeled probe.

probe with salt and ethanol. Gel filtration and commercially-available purification kits can also be used.

Testing the Labeled Probe

In this part of the exercise, you will verify that the DNA probe you prepared was adequately labeled. This will be done by making serial dilutions of your labeled DNA and labeled control DNA, then determining the smallest detectable quantify of each DNA. The overall procedure for this part of the laboratory exercise is shown in Figure 6.2.

Procedure

Dilute and Spot the Labeled DNAs

1. Thaw the probe DNA tube. Remove 1 µl of probe DNA for testing incorporation.

2. Label four, 0.5 or 1.7 ml microcentrifuge tubes with your initials and A, B, C, D, respectively. To tube "A" add 9 µl TE buffer. To B-D, add 18 µl TE buffer.

3. Then add 1 µl of probe DNA to tube "A". Mix, then transfer 2 µl of the solution from tube "A" to tube "B", mix, then transfer 2 µl of the solution from B to C, then 2 µl from C to D. Use a new pipet tip for each transfer.

4. Repeat the process above, but this time dilute the control DNA. Start with 1 µl of DNA from the control, and dilute into 4 tubes as above. Label tubes A-C, B-C, C-C, D-C (the "C" indicating "control").

5. Label the membrane with a pencil using "C" for control and "P" for probe. Label the columns A, B, C, D. Spot 1 µl of each of your dilutions (your

labeled probe and the control) on a membrane. Do the spotting as diagramed in Figure 6.2.

6. Place the membrane, DNA side down, on a piece of plastic wrap on the UV transilluminator. Irradiate for 2 minutes.

Detect the DNAs on the Membrane

All the following incubations are done at room temperature.

1. Wet the membrane with washing buffer for a few seconds, pour off the washing buffer, then incubate the membrane in blocking buffer for 5 min.

2. Drain the blocking buffer, then add 1:5000 dilution of anti-digoxigenin antibody conjugated to alkaline phosphatase in blocking buffer (1 μl antibody per 5 ml of blocking buffer). Incubate for 10 min. with gentle shaking.

3. Wash the membrane twice in washing buffer, 5 min. each time.

4. Pour off the washing buffer, incubate in detection buffer for 2 min. Then add color substrate solution (the reaction will take 15–45 min.). Compare the intensity of your spots with the control to estimate incorporation. Determine the number of ng of labeled DNA in your solution. See the tip below.

Tip

You should get a signal with at least the 1:100 dilution ("B") of your probe DNA and a signal with the 1:1000 dilution ("C") of labeled control DNA. The "B" dilution of the control contained 100 pg of DNA per μl; the "C" dilution contained 10 pg/μl; the "D" dilution contained 1 pg/μl. If the labeling reaction worked well, you would expect to get a total of about 100 ng of labeled DNA.

Calculate the amount of labeled DNA your reaction generated. Put this information in the worksheet.

References

Badenes, M. and D. Parfitt. 1994. Reducing background interference on Southern Blots probed with Nonradioactive Chemiluminescent Probes. Biotechniques 17: 622, 624.

The DIG System User's Guide for Filter Hybridization. 1995. Boehringer Mannheim, Indianapolis, IN.

DNA Probe Labeling and Testing

Name:_____   **Date** _____

Results

Yield of probe _____ Calculations:

Questions

1. Methods are available for generating single-stranded DNA for labeling. Would this DNA need to be boiled before labeling? Why or why not?

2. What was the purpose of testing the probe after labeling?

Pre-hybridization, Hybridization

Key Terms

Pre-hybridization: incubating a DNA-containing membrane in a solution designed to reduce non-specific binding of probe DNA to the membrane. The prehybridization solution you are using contains skim milk powder and detergents (SDS and N-laurlysarcosine) to prevent non-specific binding of DNAs to the membrane.

Hybridization: incubating a DNA-containing membrane in a solution containing prehybridization mix, plus a boiled, labeled probe. The salt (SSC) in the prehybridization/hybridization mix allows matching DNAs to hybridize to one another. The other components (see above) prevent non-specific binding of the probe to the membrane.

In this laboratory, the DNA probes (prepared in laboratories 5 and 6) will be incubated with the DNA transferred to the membrane in laboratory 4. Only DNA sequences that are identical or nearly identical will join together. In laboratory 8, you will then detect the binding of the probe DNA to the membrane-bound DNA.

Precautions

- SDS and N-laurylsarcosine are respiratory irritants. Wear a suitable dust mask when weighing these powders for making the stock solutions.

Materials

- Labeled probe in microcentrifuge tube with lid lock (from Lab 6)
- Membrane with DNA (from Lab 4)
- Heat-sealable bags
- Bag sealer
- Boiling water bath
- Prehybridization and Hybridization solution
- 65°C shaking water bath

Photographic Atlas Reference
Chapter 4

- Sodium hydroxide (NaOH) is a strong base. Wear appropriate protective clothing when working with NaOH.

Preparation of Solutions

Pre-hybridization, Hybridization Solution

Stock Solutions:

20X SSC — Prepared in Laboratory 4 (directions in Appendix)

10% SDS — Add 20 g of SDS to distilled water to a final volume of 200 ml. Wear a mask when weighing SDS, since SDS dust is a respiratory irritant. Wipe up any spilled powder with a wet cloth.

10% N-laurylsarcosine — Add 20 g N-laurylsarcosine to distilled water to a final volume of 200 ml. Wear a mask when weighing out the powder. Wipe up any spilled powder with a wet cloth.

Preparing the Working (1X Concentration) Pre-hybridization/Hybridization Solution

For a 500 ml solution, add

- 125 ml 20X SSC
- 360 ml distilled water
- 10 g powdered skim milk
- 5 ml 10% N-laurylsarcosine
- 1.0 ml 10% SDS.

Be sure to add the SDS last. If the SDS is mixed directly into the 20X SSC solution, it will form an insoluble precipitate. Hybridization solution is prehybridization solution plus labeled, boiled probe.

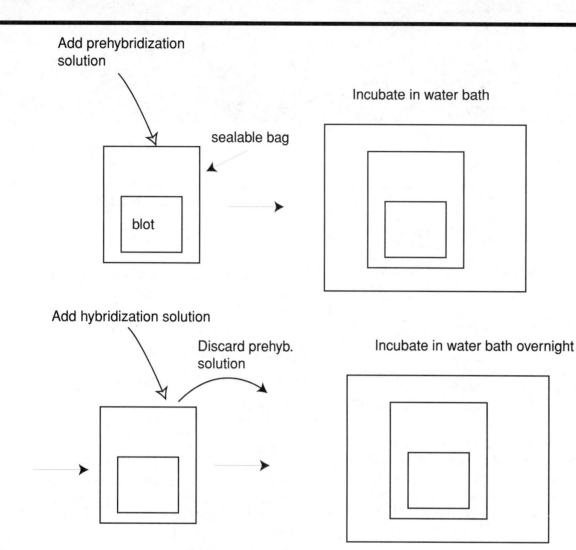

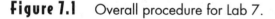

Figure 7.1 Overall procedure for Lab 7.

Procedure for Pre-hybridization, Hybridization (Figure 7.1):

Prehybridization

1. Put your membrane in a heat-sealable pouch. Add 11 ml of hybridization solution to bag (formula is 0.2 ml/ cm²; 11 ml is sufficient for a membrane with a 56 cm² area (7 X 8 cm gel). Adjust volume as required for other membrane sizes).

2. Seal the bag with a heat sealer, avoiding bubbles. The easiest way to remove bubbles is to place the bag on the lab bench with the opening away from you. Raise the open end of the bag slightly. Starting from the bottom of the bag, roll the bubbles up to the top with a pipet. When you near the top, slightly release your hold on the pipet so that the trapped liquid can flow back down into the bag, but that the bubbles remain trapped (Figure 7.2). Seal the bag below the level of the bubbles.

3. Quickly check for leaks by turning the bag so the sealed side is down. If the bag leaks, reseal. When the bag is leak-free, place the sealed bag into a plastic container in a 65°C shaking water bath. Prehybridize at 65°C for 30 min. to 1 hour.

Preparation for Hybridization

1. Prepare a boiling water bath. When your blot is within 10 minutes of the end of the prehybridization,

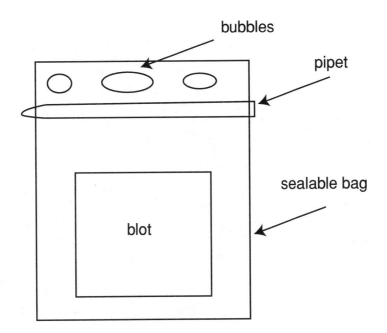

Figure 7.2 Removing bubbles from the blot.

boil the probe solution in a microcentrifuge tube (with a lid-lock) for 3 min. Then place the boiled probe on ice immediately. (Boiling the probe makes it single-stranded; the probe must be single-stranded in order to bond with the single-stranded DNA on the membrane.)

2. Pipet or pour 11 ml of prehybridization buffer into a sterile tube with a watertight lid (plastic centrifuge tubes work well).

3. Add the boiled probe solution to the tube with prehybridization buffer. Mix rapidly.

4. Immediately take the prehybridizing blot out of the incubator. Pour the old prehybridization solution out of the bag. Immediately add the hybridization solution (prehybridization buffer plus the boiled probe). **It is critical that the membrane does not dry out between pouring off the prehybridization solution and adding the hybridization solution.** Reseal the bag, avoiding bubbles as described above.

5. Hybridize overnight at 65°C (hybridization for two days normally doesn't cause problems).

If time remains in your laboratory period, your instructor may want you to prepare the solutions for the detection reaction, which will be used in the next laboratory session.

Tip

If your labeling reaction did not work, you can use half of the labeling reaction from a classmate who did have a successful reaction. If you don't need to share, use the entire labeled probe for the hybridization reaction.

References

Badenes, M. and D. Parfitt. 1994. Reducing background interference on Southern Blots probed with Nonradioactive Chemiluminescent Probes. Biotechniques 17: 622, 624

The DIG System User's Guide for Filter Hybridization. 1995. Boehringer Mannheim, Indianapolis, IN.

Pre-hybridization, Hybridization

Name:_____   **Date** _____

Questions

1. With some protocols, the labeled probe is not boiled, but sodium hydroxide is added to the probe before the hybridization reaction. What must the sodium hydroxide do?

2. If the probe sequences do not exactly match the target sequences, the hybridization reaction is typically done at a lower temperature. (This could occur for example, if you using probe DNA from one species to identify similar DNA sequences in another species.) Explain.

Detecting λ DNA on the Membrane

Key Term

Antibody: A protein, produced by the immune system of animals, that specifically recognizes another molecule. In this laboratory exercise, we are using an antibody that specifically recognizes and binds to digoxigenin. DNA on the membrane that binds to the digoxigenin-labeled probe DNA, will be bound by the antibody and detected by an enzyme carried on the antibody.

This is the final laboratory in the series that you started with laboratory 3. You will be using the membrane containing λ and λgt10 DNA (prepared in lab 4) and the labeled λ DNA fragment (isolated in laboratory 5 and labeled in laboratory 6). In this lab, you will be able to determine if λgt10 is missing a section of DNA that is present in λ. If the λ probe DNA is not present in λgt10, you would expect to see a band (or bands) only in the λ lanes; if the DNA is present in λgt10 you would expect to see bands in all the lanes.

Materials

- Sealable plastic container
- Incubator set at 65°C
- Wash solution I
- Wash solution II
- Washing buffer
- Blocking buffer
- Detection buffer
- Anti-digoxigenin Antibody
- Color substrate buffer
- Distilled water

Photographic Atlas Reference
Chapter 4

Precautions

Follow the general guideline of avoiding contact with any laboratory chemicals.

Preparation of Solutions

Wash solutions, Washing buffer, Blocking solution, Detection buffer, and Color substrate buffer are prepared as described in the Appendix. Except for the Wash solutions, all these reagents were prepared in laboratory 6.

Procedures

(Note: These volumes are for a 56 cm² blot. Adjust the volumes proportionately for larger or smaller blots.)

Washing Off Excess Probe

(Note: It is critical that the membrane does not dry out between any of the washing steps.)

1. Open the bag and pour off hybridization solution. Transfer the membrane to a small, sealable plastic container. Immediately add 25 ml Wash solution I to blot, incubate 5 min. @ 65°C in a shaking incubator. Pour off wash solution.

2. Add 25 ml of Wash solution II, incubate 15 min. @ 65°C in a shaking incubator. Pour off Wash solution II and immediately transfer the blot to a different, clean container. Be sure you proceed immediately with the next step.

All the following incubations are done at room temperature with gentle shaking, either by hand or on a rotary or tilting platform. The solutions are the same as those used in laboratory 6 for determining the efficiency of probe labeling (see Appendix for directions on preparing the solutions, if necessary).

1. Immediately add 10 ml of Washing buffer, incubate for 1 min at room temperature.

2. Pour off Washing buffer, add 30 ml of Blocking solution, incubate for 30–60 minutes at room temperature.

3. Pour off Blocking solution, add 20 ml fresh Blocking solution containing 4 µl anti-digoxigenin antibody. Incubate at room temperature for 15–30 min (this incubation can be extended to at least two hours without causing problems).

4. Discard the Blocking solution containing the antibody. Add 50 ml of Washing buffer. Incubate for 10 min. Discard Washing buffer, add 50 ml of fresh Washing buffer, incubate for another 10 minutes.

5. Pour off the Washing buffer, add 10 ml of detection solution, incubate for 2 minutes.

6. Then add just enough Color Substrate Buffer or Western Blue® (Promega Biotech) to cover membrane (~5 ml). Incubate *without shaking*, in the dark, until color appears (make sure membrane is DNA-side up). This should take 15 to 30 minutes (as long as 18–20 hours if probe labeling was poor). Normally the blot can sit in the Color Substrate Buffer overnight (in the dark) when covered to prevent evaporation.

7. When color is sufficiently dark, pour off detection solution, rinse with distilled or deionized water or TE. Allow blot to dry in your drawer (i.e., in the dark) on a paper towel. Answer the questions in the back of the lab manual.

Tips

Although this procedure normally works very well, because of the number of steps involved, students occasionally experience problems. The problems normally fall into two categories: no signal or too high background staining. Some common causes of those problems are described below.

No Signal (no bands on the gel, no purple color)

1. Probe was not properly labeled (this was tested in Lab 6).

2. Probe was not boiled before it was added to the prehybridization solution.

3. DNA did not transfer to the membrane. This should have been determined in laboratory 4 — the loading dyes should have been visible on the membrane and during the UV-fixation step, faint orange DNA bands should have been visible on the membrane.

High Background (purple color throughout the membrane)

1. The blocking solution was not prepared properly or the blocking reaction was not performed for a long enough time.

2. The membrane was allowed to dry out between steps.

3. The membrane was not kept in the dark during the incubation with Western Blue® or the color substrate solution.

4. The dried membrane was exposed to bright light.

5. Too much anti-digoxigenin antibody was used.

6. The reaction was allowed to go on too long.

Reference

The DIG System User's Guide for Filter Hybridization. 1995. Boehringer Mannheim, Indianapolis, IN.

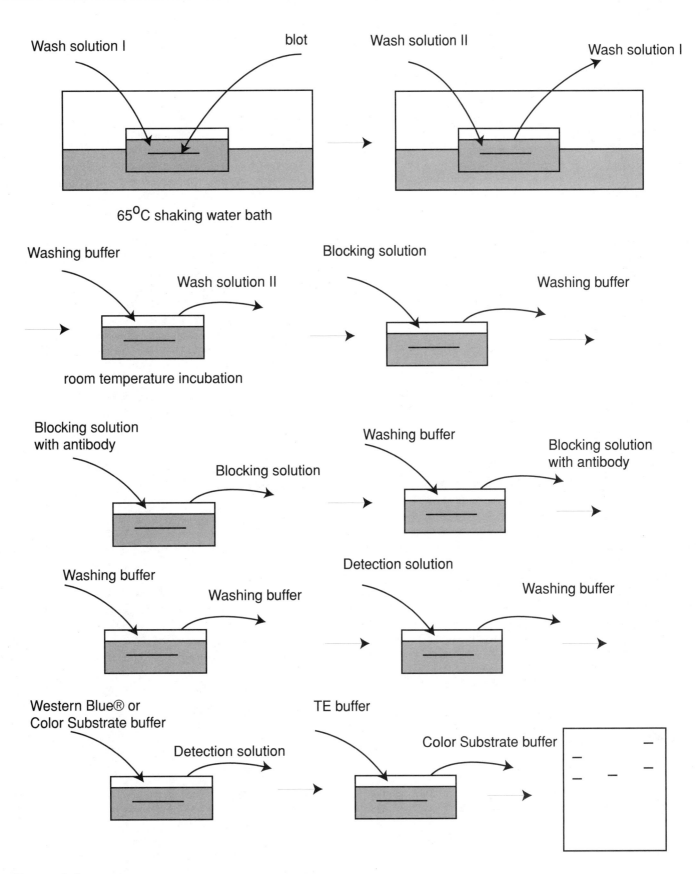

Wash solution I   blot

Wash solution II

Wash solution I

65°C shaking water bath

Washing buffer

Wash solution II

room temperature incubation

Blocking solution

Washing buffer

Blocking solution
with antibody

Blocking solution

Washing buffer

Blocking solution
with antibody

Washing buffer

Washing buffer

Detection solution

Washing buffer

Western Blue® or
Color Substrate buffer

Detection solution

TE buffer

Color Substrate buffer

Figure 8.1   Overall procedure for Laboratory 8.

Detecting λ DNA on the Membrane

Name: _____   Date _____

Results

Questions

1. Explain the pattern of labeled bands you observed for the λ and λgt10 DNAs.

2. What result would you have expected to see if the entire λ genome was used for labeling, not just the 2322 and 2027 bp bands?

Polymerase Chain Reaction (PCR)

Key Terms

Polymerase Chain Reaction (PCR): A method of exponentially amplifying a specific sequence of DNA. PCR requires template DNA (in this case from your cheek cells), primers, dNTPs, a heat-stable polymerase, and the appropriate buffer.

Primers: Short, single-stranded DNAs; also called oligonucleotides. For PCR, primers are typically 20–30 nucleotides long.

Taq DNA Polymerase: a heat-stable enzyme that catalyzes the synthesis of DNA.

Materials for Option A

- Buccal Swab kit from Epicentre Technologies, Inc, which includes:
  - A. Nylon Buccal brushes (one per student)
  - B. Tube of DNA extraction solution (one per student)
- Vortex mixer
- 60°C water bath
- 98°C water bath
- 1 PCR reaction tube (thin-walled 0.2 ml or 0.5 ml depending on the machine you have)
- ApoC2 primer 1 (25 pmoles/μl) 5'CATAGC-GAGACTCCATCTCC
- ApoC2 primer 2 (25 pmoles/μl) 5'GGGA-GAGGGCAAAGATCGAT
- dNTP mix (10 mM each nucleotide; Sigma or other suppliers)
- 10X PCR buffer with 25 mM $MgCl_2$
- *Taq* DNA polymerase
- Sterile, deionized water
- PCR thermocycler programmed for 94°C, 55°C, 72°C for 35 cycles

Photographic Atlas Reference
Chapter 5

The polymerase chain reaction (PCR) is a widely used technique in molecular biology. PCR is a method for amplifying a section of DNA bracketed by two primer sequences. In this lab, you will be amplifying a fragment of DNA from your cheek cells.

The section of DNA you will be amplifying is the APOC2 locus which is located on chromosome 19. There are at least 11 alleles for this locus, which vary in the number of two-base repeats (up to a maximum of 30 repeats). A person can have two identical alleles (homozygous) or two different alleles (heterozygous). The repeated sequence is in an intron in the apolipoprotein C2 gene (involved in cholesterol transport). This type of locus is variously called a VNTR (for variable numbers of tandem repeats), STR (for short tandem repeat), or minisatellite.

These types of polymorphic loci are particularly useful for distinguishing individuals in forensics and paternity cases. When several different polymorphic loci are simultaneously tested, the odds for or against a match quickly become astronomical, since the odds are the product of the odds for each individual test.

In the lab exercise, the cheek cells are placed in a hypotonic solution in a microcentrifuge tube that contains a resin that protects DNA and binds PCR inhibitors. The solution is then heated. The hypotonic solution (in concert with high temperature) breaks open the cells and releases the DNA. Then PCR reagents are added to allow a specific DNA segment to be amplified.

Precautions

Be sure to discard all materials that have contacted the cheek cells (inoculating loop, boiled tubes) into a biohazard waste container, followed by autoclaving. Follow precautions described in laboratory 2 for agarose gel electrophoresis.

Option A: Extraction Using a Kit (See Figure 9.1A)

Procedure: Adapted from the protocol for the Buccal Swab Kit (Epicentre Technologies, Madison, WI).

1. Thaw one tube of DNA extraction solution per person.

2. Thoroughly rinse out your mouth with water two times.

3. Harvest cells by brushing the inside of the cheek with the nylon swab. Do about 20 strokes with the swab on each cheek.

4. Transfer the brush (with cells) to the DNA Extraction Solution. Twirl the brush in the solution 5 or 6 times. Rotate the brush when removing it from the tube, to squeeze most of the liquid out. Discard the brush in the biohazard waste.

5. Screw the cap on, vortex mix for 10 seconds. Place the tube in a water bath at 60°C for 30 minutes. When the incubation is done, vortex mix the tube contents for 15 seconds.

6. Transfer the tube to a 98°C water bath or incubator (a boiling water bath works OK); hold at 98°C for 8 minutes. After the incubation, vortex mix for 15 seconds.

7. Return the tube to the 98°C water bath or incubator; hold for another 8 minutes. After the incubation, vortex mix for 15 seconds. Chill the tube on ice for 1 minute.

8. Centrifuge at maximum speed (12,000 X g) at 4°C for 5 minutes.

9. Either use the supernatant (which contains the DNA) immediately for the PCR reaction in part B, below, or transfer the supernatant to a clean tube without transferring any of the beads. The DNA can be stored at −20°C until the next laboratory period. Otherwise, proceed directly to the PCR amplification in Section B, below.

Preparing the PCR reaction

1. Label the PCR tube on the top and sides with your initials. Be careful to avoid creating bubbles in the following step. The PCR buffer normally contains a detergent. Vigorous pipetting up and down may create bubbles in the tube. Keep your pipet tip below the surface of the liquid, and avoid pushing air out of your pipet tip as you mix the contents of the tube. If bubbles form, centrifuge the tube briefly prior to placing it in the thermocycler.

2. To each tube, add
   - 1 μl of primer 1 (25 pmoles)
   - 1 μl of primer 2 (25 pmoles)
   - 1 μl of dNTP mix (10 mM solution of each nucleotide)
   - 5 μl of 10X PCR buffer with $MgCl_2$
   - 36 μl of sterile, distilled water
   - 5 μl of DNA solution from your cheek cells
   - 0.5 μl of *Taq* DNA polymerase (5 units/μl).

   Overlay with mineral oil if required. Keep on ice prior to loading into the thermocycler.

3. When everyone is ready, turn the thermocycler on. Pause the thermocycler when the temperature reaches 94°C. Each person should then quickly place his or her tube in the thermocycler. When all the tubes are in the thermocycler, start the program. The thermocycler should be set to run at 94°C for 5 min (one cycle), then 35 cycles of 94°C for 1 min., 55°C for 1 min., 72°C for one minute.

4. While the thermocycler is running, prepare a 2.5% agarose gel, 9 people per gel, two 10 well combs per gel.

Next Laboratory Session

- Run the agarose gel.
- Each person should load their samples into the same lane in the top and bottom parts of the gel.
- Top row: PCR products: 10 μl PCR reaction, 2 μl 6X loading buffer.
- Bottom row: Repeat (this increases the chance of seeing your PCR product, if something went wrong with loading the first sample).
- Run the gel, then photograph the gel, and analyze the data

Option B: Laboratory-Prepared Solutions (See Figure 9.1B)

Procedure

A. Preparing DNA for Amplification

Rinse your mouth out prior to starting the exercise.

1. Gently scrape the inside of your cheek with the sterile inoculating loop. Twirl the loop in the

A) PCR procedure with kit

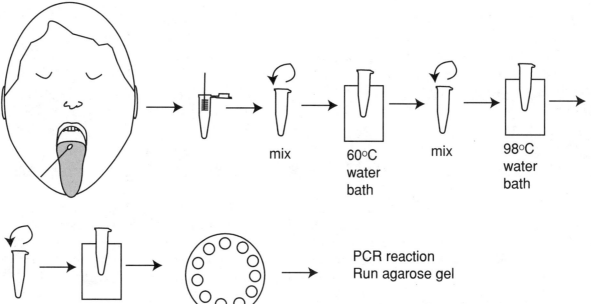

mix

60°C
water
bath

mix

98°C
water
bath

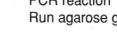

mix

98°C
water
bath

centrifuge

PCR reaction
Run agarose gel

B) PCR procedure without kit

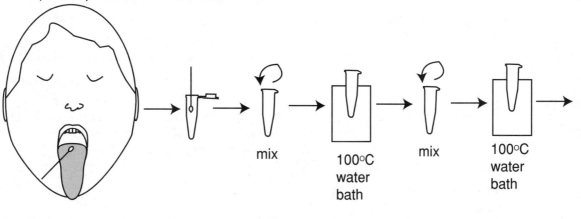

mix

100°C
water
bath

mix

100°C
water
bath

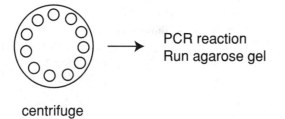

centrifuge

PCR reaction
Run agarose gel

Figure 9.1   Overall procedure for Laboratory 9 A: Kit method B: Lab-prepared solutions.

Materials for Option B

- Two sterile, 10 μl inoculating loops
- Lid-locks for microcentrifuge tubes
- One, 1.7 ml microcentrifuge tube with 500 μl of water with 5% (weight/volume) Chelex® resin
- One PCR reaction tube (thin-walled 0.2 ml or 0.5 ml depending on the machine you have)
- ApoC2 primer 1 (25 pmoles/μl) 5'CATAGCGA-GACTCCATCTCC
- ApoC2 primer 2 (25 pmoles/μl) 5'GGGA-GAGGGCAAAGATCGAT

- dNTP mix (10 mM each nucleotide)
- 10X PCR buffer with 25 mM $MgCl_2$
- Sterile, deionized water
- *Taq* DNA polymerase (typically 5 units/ml)
- Vortex mixer
- Microcentrifuge
- Water bath set at 98°C or boiling
- PCR thermocycler programmed for 94°C, 55°C, 72°C for 35 cycles. A thermocycler with a heated lid is preferred. If none is available, PCR grade mineral oil will also be required.

Chelex solution for 5 to 10 seconds to dislodge cells. Repeat with a second sterile inoculating loop. Label the tube on the top and side with your initials. Discard the loops in the biohazard waste bag.

2. Put a lid-lock on the tube to keep the cap from popping off. Boil for 8 minutes. Vortex mix the sample for 15 seconds.

3. Boil the sample again for 8 minutes. Vortex mix again for 15 seconds.

4. Spin in a microcentrifuge (be sure the tubes are balanced) for 3 minutes (11,000 × g, ~13,000-14,000 rpm) at maximum speed. Place the 1.7 ml tube on ice.

B. Preparing the PCR reaction

1. Label the PCR tube on the top and sides with your initials.

2. To each tube, add:
   - 1 μl of primer 1 (25 pmoles)
   - 1 μl of primer 2 (25 pmoles)
   - 1 μl of dNTP mix (10 mM each dNTP)
   - 5 μl of 10X PCR buffer with $MgCl_2$
   - 31 μl of sterile, distilled water
   - 10 μl of DNA solution from your cheek cells
   - 0.5 μl of *Taq* DNA polymerase (5 units/μl).

   Overlay with mineral oil if required. Allow to sit on ice prior to loading into the thermocycler.

3. When everyone is ready, turn the thermocycler on. Pause the thermocycler when the temperature reaches 94°C. Each person should then immediately

place his or her tube in the thermocycler. When all the tubes are in the thermocycler, start the machine running. The thermocycler should be programmed to run at 94°C for 5 min (one cycle), then 35 cycles of 94°C for 1 min., 55°C for 1 min., 72°C for 1 min.

4. While the thermocycler is running, prepare a 2.5% agarose gel, 9 people per gel, two 10 well combs per gel.

Next Laboratory Session:

- Run the agarose gel.
- Each person should load their samples into the same lane in the top and bottom parts of the gel.
- Top row: PCR products: 10 μl PCR reaction, 2 μl 6X loading buffer.
- Bottom row: Repeat (this increases the chance of seeing your PCR product, if something went wrong with loading the first sample).
- Then run, photograph and analyze the gel.

References

Roth, B., M. Thompson, and R. Hallick. 1997. DNA fingerprinting in a High School Research-Based Science Course. The American Biology Teacher 59(1): 48–51.

Campbell, A. M., J. H. Williamson, D. Padula, and S. Sundby. 1997. Use of PCR and a Single Hair to Produce a "DNA Fingerprint". The American Biology Teacher 59(3): 172–178.

Polymerase Chain Reaction (PCR)

Name: _____   **Date** _____

Results

Determine the size of the PCR product(s) from your amplification reaction. Plot on semi-log paper (next page). Is possible to distinguish individuals in your gel using this locus?

Questions

1. What result would you expect if you added no primers to the PCR reaction?

2. Why is an enzyme from a thermophilic bacterium used for the PCR reaction?

3. What are some potential concerns about using this technique for forensic identification or for identification of pathogens?

4. In spite of the concerns listed in "3" above, why is this technique so widely used and so important in modern biological research.

III
RNA Isolation & Analysis Techniques

The next set of laboratory exercises will teach you a method for isolating RNA and two methods for analyzing the isolated RNA. In the first exercise (Lab 10) you will use a method of RNA isolation that prevents the degradation of RNA while simultaneously separating the RNA from DNA and proteins, based on differences in the chemical properties of those bio-molecules. In Laboratory Exercise 11, you will use a rapid PCR based method (RT-PCR) for determining whether a mRNA is present in the RNA isolated in the previous exercise. In Laboratory Exercise 12, you will use Northern blotting to determine the size of a mRNA.

Key Terms

mRNA: messenger RNA; unstable RNAs that are translated to produce proteins in the cell. Isolation and analysis of mRNA is one of the most common methods for determining when and to what extent a gene is expressed.

Materials

For yeast growth:

- Fresh baker's yeast (a grocery store variety works fine)
- Sabouraud Dextrose Agar
- Sterile petri plates
- Shaking incubator at room temperature
- 1 test tube with 5 ml of YPD medium per student or group (sterile)

RNA isolation reagents:

- Sterile, plastic, RNase-free microcentrifuge tubes
- Refrigerated centrifuge
- Tri-Reagent (for mRNA isolation)
- 1-bromo-3-chloropropane (BCP) (a less-toxic compound than chloroform — chloroform can be substituted)
- Isopropanol
- Acid-washed glass beads, 425-600 micron (Sigma Cat #G 8772) or 710-1180 microns (Sigma Cat #G9393)
- Vortex mixer
- p20, p200, p1000 pipettors, RNase-free pipet tips
- DMPC or DEPC-treated water
- Parafilm®
- 75% Ethanol, RNase-free
- If available: UV spectrophotometer, quartz cuvettes

Photographic Atlas Reference
Chapter 6

In this exercise you will be isolating RNA from the yeast *Saccharomyces cerevisiae*. The isolated RNA will then be used for analysis of expression of the phosphoglycerate kinase gene (*PGK*) by Reverse-Transcriptase PCR (RT-PCR) and northern Blotting.

Precautions

RNA isolation involves handling a number of potentially toxic compounds. The Tri-reagent, 1-bromo-3-Chloropropane, isopropanol, and DMPC or DEPC are all potentially hazardous. Wear protective clothing and follow institutional guidelines and the manufacturer's MSDS for details on appropriate precautions.

Preparation of Media and Solutions
Sabouraud Dextrose Agar

Follow directions from manufacturer, but typically 65 grams of medium are added to distilled water to make 1 liter of medium. Autoclave at 121°C for 15 minutes. Pour into plates, allow to solidify, refrigerate until needed.

YPD Broth Medium (Sigma)

Add 50 grams to make 1 liter of medium. Dispense in 5 ml aliquots into capped test tubes. Autoclave at 121°C for 15 min.

DMPC-treated Water

In a working chemical fume hood, add 1 ml of Dimethyl pyrocarbonate (DMPC Sigma Cat #D5520) to 1 liter of distilled or deionized water in an autoclavable bottle. Cap tightly, mix thoroughly. Allow to sit at room temperature for 30 minutes. Then loosen cap, autoclave for 15 min. at 121°C. Sterile, RNase-free water can also be purchased commercially.

Procedures

Yeast Growth
Pre-lab:

Start with baker's yeast (*Saccharomyces cerevisiae*, from the grocery store).

Add a small amount of dried yeast to a tube containing sterile YPD medium. Mix to disperse. Then streak out the yeast suspension to isolate colonies on Sabouraud Dextrose Agar (this yeast preparation may contain bacteria, so the yeast needs to be isolated). Allow to grow two days at room temperature. Restreak from a single colony to a new Sabouraud dextrose agar plate. Allow to grow two days at room temperature. After the yeast has grown for two days on the plate, the plate can be refrigerated for several weeks and the cells will still be viable.

Day Prior to Lab

Inoculate an isolated colony from the plate into a test tube containing 5 ml sterile YPD broth. Allow to grow overnight with shaking at 25°C (room temperature).

mRNA isolation (Figure 10.1)

(Based on the Tri-Reagent™ protocol from the Molecular Research Center, Cincinnati, OH)

1. Pour ~1.0 ml of the yeast culture into a sterile 1.7 ml RNase-free microcentrifuge tube. Place the tube into the microcentrifuge. *Be sure the centrifuge is balanced with an equal number of tubes on either side of the rotor before the centrifuge is started. Use a tube filled with water to balance the centrifuge, if necessary.* Pellet the cells by centrifugation at 3000 rpm for 2 min. in a centrifuge. Pour off the supernatant.

2. Resuspend the cells by vortex mixing the tube for 10 sec.

3. Add approximately 400 μl of acid-washed glass beads. Add 1 ml Tri-reagent™. Close cap and wrap Parafilm® around the rim of the tube to prevent the lysis reagent from escaping if the tube pops open. Lyse cells by vigorous vortex mixing for 2 min.

4. Remove the Parafilm®. Centrifuge the sample at maximum speed (12,000 X g) for 10 min. at 4°C. Transfer the liquid to a new tube by pipetting out the supernatant.

5. Allow the samples to stand at room temperature for 5 minutes.

6. Add 0.1 ml of 1-bromo-3-chloropropane (BCP) (0.2 ml of chloroform can be substituted). Close cap tightly, shake vigorously for 15 seconds (don't vortex mix). After mixing, let samples sit at room temperature for 10 minutes.

7. Centrifuge at maximum speed (about 12,000 X g) for 15 min. at 4°C. The tube should have two distinct layers: A clear aqueous phase on top containing the RNA and a reddish lower organic layer containing DNA, protein, and other cell debris. There also may be a thin whitish interface between the two phases: this contains denatured proteins.

8. Transfer ONLY the upper aqueous phase (clear) to a new, sterile, RNase-free 1.7 ml microcentrifuge tube with a pipettor (~400 μl). Put the tip along the side of the tube and slowly suck up the liquid. Take care to avoid pipetting up any of the white interface or the red organic phase. It is better to leave some of the upper aqueous layer behind, rather than contaminating your preparation with the other material.

9. Optional, if time allows. Add 1/10 volume of isopropanol (~ 40 μl), mix by inversion. Allow tube to sit at room temperature for 5 min. Centrifuge at maximum speed (12,000 X g) for 10 minutes at 4°C. Pipet supernatant to a new tube. (This procedure is designed to remove residual DNA: it is particularly useful if the isolated RNA will be used for RT-PCR without DNase treatment.)

10. Add 0.46 ml of isopropanol if you did Step 9 (or 0.5 ml isopropanol if you did not do Step 9). Mix thoroughly by inversion. Allow the sample to sit for 5 minutes at room temperature. Centrifuge at maximum speed (12,000 X g) at 4°C for 10 minutes. The RNA should pellet to the bottom of the tube.

11. Pour off the supernatant, then add 1 ml of RNase-free 75% ethanol. Vortex the pellet to dislodge it from the bottom of the tube. (The RNA sample at this point can be stored at −20°C. Immediately before the RNA will be used, complete the RNA isolation by doing steps 12 and 13, below. If you complete the RNA isolation (through Step 13), the RNA should be stored at −70°C.)

12. Centrifuge the sample at maximum speed (12,000 X g) for 5 min. Pour off the supernatant. Spin the samples for 5 seconds in the microcentrifuge, pipet up remaining liquid. Allow the RNA pellet to air dry with the lid open for 10 min.

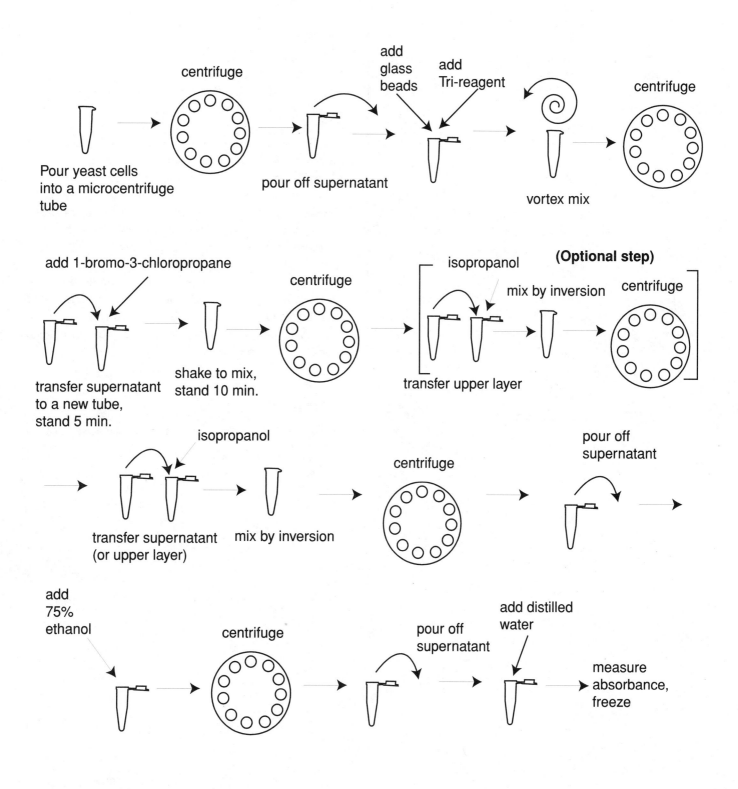

Figure 10.1 Overall procedure for Lab 10.

13. Resuspend the RNA in 20 µl sterile, deionized, RNase-free water. Transfer 2 µl to a separate tube. Store the remaining RNA at −70°C if possible (otherwise, store at −20°C).

14. Measure the absorbance of the 2 µl RNA sample by spectrophotometry at 260 and 280 nm if a UV spectrophotometer is available. Use the following formula to calculate the concentration of RNA (in µg/µl of your RNA solution):

$$\text{RNA conc. (µg/µl)} = \frac{OD_{260} \times \text{dilution factor} \times 40}{1000}$$

- OD_{260} means the optical density reading at 260nm
- *dilution factor* means how many-fold the sample was diluted. For example, 1 µl in 100 µl would be a 100-fold dilution (and would have a dilution factor of 100)
- 40 this is the extinction coefficient for RNA.

So for example, if you were working with a total volume of 100 µl in the cuvette, which included 5 µl of your RNA solution, and you got an OD reading at 260nm of 0.1, then you would have:

$$\text{RNA conc.} = \frac{0.1 \ (OD_{260}) \times 20 \times 40}{1000} =$$

0.08 µg/µl in the RNA solution that you isolated.

- The 260/280 O.D. ratios give an indication of purity. A ratio of 2.0 is pure RNA. Lower ratios (especially below 1.6) indicate contamination with proteins, or chemicals used in the isolation.

Tips

Working with RNA is much more difficult than working with DNA. RNases are widespread in the environment and extremely resistant to most treatments that inactivate enzymes. Therefore, any work with RNA has to be conducted with extreme care to ensure that no RNases come in contact with isolated RNA. All work with RNA should be done aseptically. Gloves should always be worn and changed regularly. Whenever possible, certified RNase-free, disposable, sterile, plastic tubes, pipets, etc. should be used. Any solutions to be used for RNA work should be treated with Dimethylpyrocarbonate (DMPC) (see precautions above). Great care should be taken to ensure that treated solutions do not become contaminated with bacteria (which are rich in RNases) or contaminated with solutions or labware that have not been treated to remove RNases.

Reference

Tri-reagent Protocol Update. 1998. Molecular Research Center, Cincinnati, OH.

mRNA Isolation

Name:_____   **Date** _____

Results

Amount of RNA isolated _____Calculations:

Questions

1. RNA degradation by RNases is a much greater problem than DNA degradation by DNases. Based on the biological function of RNA and DNA molecules, why do you think this is so?

2. Most methods of RNA isolation include one or more protein denaturants in the lysis buffer. Why do you think this is so?

3. Sometimes, when RNA is isolated for RT-PCR, the RNA solution is treated with DNase. Why might this be more important than if the RNA was only going to be used for northern hybridization?

Reverse Trancriptase PCR (RT-PCR)

Key Terms

RT-PCR: Reverse-transcriptase polymerase chain reaction. A rapid RNA analysis method. The procedure involves reverse transcription of mRNA into DNA, followed by PCR amplification.

Reverse transcriptase: an enzyme, typically isolated from eukaryotic retroviruses, that catalyzes the synthesis of DNA from a mRNA template.

Photographic Atlas Reference
Chapter 6

In this laboratory exercise, you will be using Reverse-Transcriptase (RT) PCR to analyze the RNA prepared in the previous laboratory exercise. RT-PCR is a very rapid, semi-quantitative way of assaying transcription of a specific gene. RT-PCR is performed in two steps. First, the reverse trancriptase enzyme uses RNA as a template to synthesize DNA. Then, PCR is used to amplify the DNA to give a detectable signal.

In this exercise, you will be determining whether the *PGK* mRNA is present in yeast cells grown overnight in YPD medium. (The PGK protein is essential for glucose metabolism in the yeast cell.)

Materials

- *PGK* up primer (25 pmoles/µl): 5'-TGCCCCAGGTTCCGTTATTTTGTT
- *PGK* down primer (25 pmoles/µl): 5'-TTCTGGACCATTGTCCAACCCTTG
- Omniscript™ RT-PCR kit (Qiagen: includes 10X RT buffer, 5mM dNTP mix, RNase-free water, and the reverse trancriptase enzyme)
- Placental RNase inhibitor (recombinant, Promega Biotech)
- *Taq* polymerase, buffer, nucleotides (see Lab 9)
- Gloves
- p20, p200 pipettors, RNase-free tips
- Agarose gel electrophoresis equipment and supplies (see Lab 2)

Precautions

Be certain to wear gloves and to follow the other procedures described under "Tips" in the mRNA isolation exercise (Lab 10) in order to prevent RNA degradation.

Procedures (Figure 11.1)

RT-PCR reaction. The following directions are adapted from the Qiagen Omniscript™ kit. Follow the appropriate directions from the manufacturer if you are using a different product. Each person or each lab group should run two reactions — one reaction with reverse transcriptase added, the other reaction without reverse trancriptase.

1. In this step, you will use reverse transcriptase to synthesize DNA from *PGK* mRNA. Add a solution containing 1 µg of purified RNA to a sterile, RNase-free, 0.5 ml or 1.7 ml microcentrifuge tube. (This may require diluting the RNA isolated in the previous exercise; if you do not have an UV spectrophotometer for calculating RNA concentration, add 0.5 µl of the RNA isolated in the previous laboratory exercise.) Put the RNA on ice, then add (**in order**): 2 µl of 10X RT buffer, 0.5 µl (20 units) placental RNase inhibitor, 2 µl of 5 mM dNTP mix, 1 µl of *PGK* down primer, sterile, deionized, RNase-free water to a final volume of 19 µl. Then add 1 µl of Omniscript™ reverse transcriptase. The amount of water you add will vary, depending on the amount of RNA solution you added. Use the list below to make sure you added the proper volume of all the solutions.

   - ___ µl RNA solution
   - ___ RNase-free water
   - 2 µl RT buffer

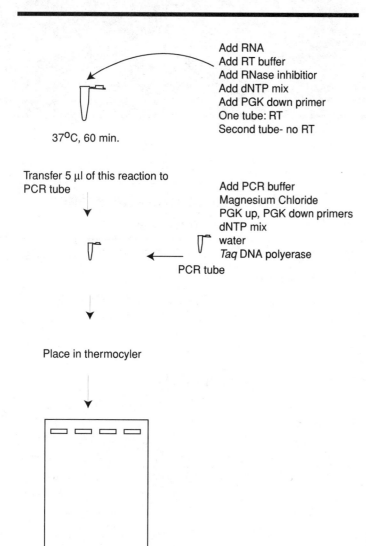

Add RNA
Add RT buffer
Add RNase inhibitior
Add dNTP mix
Add PGK down primer
One tube: RT
Second tube- no RT

37°C, 60 min.

Transfer 5 μl of this reaction to
PCR tube

Add PCR buffer
Magnesium Chloride
PGK up, PGK down primers
dNTP mix
water
Taq DNA polyerase

PCR tube

Place in thermocyler

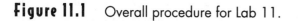

Run agarose gel

Figure 11.1 Overall procedure for Lab 11.

- 0.5 μl RNase inhibitor
- 2 μl dNTP mix (5 mM each dNTP)
- 1 μl *PGK* down primer
- 1 μl Omniscript™ Reverse transcriptase (20 μl total volume)

In a second tube, add together all the ingredients listed above except the reverse transcriptase. Use the list below to ensure you add all the reagents.

- ___ μl RNA solution
- ___ RNase-free water
- 2 μl RT buffer

- 0.5 μl RNase inhibitor
- 2 μl dNTP mix (5 mM each dNTP)
- 1 μl *PGK* down primer (20 μl total volume)

2. When you have finished adding all the reagents, incubate the solutions at 37°C for 60 minutes. Place the tubes on ice.

3. Now you will use PCR to amplify any *PGK* DNA produced in the previous steps. Set up the PCR reaction. Label one tube "No", the other tube "RT" and put your initials on both tubes. In each of the two PCR tubes, mix together:

- 5 μl 10X PCR buffer with MgCl$_2$
- 1 μl *PGK* down primer
- 1 μl *PGK* up primer
- 1 μl 10 mM dNTP mix
- 36.5 μl sterile, distilled or deionized water
- 0.5 μl *Taq* DNA polymerase

Then add 5 μl of the solution from the reverse trancriptase reaction or no reverse transcriptase reaction to the appropriate tube. Place the tubes on ice. Add one drop of mineral oil if you will be using a thermocycler without a heated lid.

4. When everyone in class is ready, start the thermocycler (it should be set for 30 cycles of 94°C for 30 seconds, 55°C for 30 seconds, 72°C for 1 minute; the machine should also be set to cool to 4°C after the reactions are finished). When the thermocycler reaches 94°C, pause the machine. Allow everyone to load his or her samples. Then restart the machine to begin the cycles.

5. The next class period, analyze the RT-PCR reaction by agarose gel electrophoresis. Prepare a 1.5% TBE-agarose gel with ethidium bromide. Run 10 μl (plus 2 μl loading buffer) of the "no RT" and "RT" added samples in adjacent lanes (e.g., load the gel with a marker DNA, then a "no RT" sample, then a sample with "RT" added, then a "no RT" sample, etc.)

Reverse Transcriptase PCR (RT-PCR)

Name:_____   Date _____

Results

Questions

1. One common control for RT-PCR is to run a reaction with RNA isolated from an organism, and primers that amplify a mRNA known to be present in that RNA mixture. What is the purpose of that control?

2. Frequently a competitor molecule is added to the RT-PCR reaction for quantification of the amount of specific mRNA present. What must be true of this competitor molecule?

3. The primers used for RT-PCR in higher eukaryotes typically span exon boundaries. Why would that be important for establishing whether a particular mRNA was expressed in a particular tissue?

Photographic Atlas Reference
Chapter 6

Key Term

Northern hybridization: A procedure where RNA, fractionated on an agarose gel, is transferred to a membrane and hybridized to a labeled probe. This technique is used to determine the size of mRNAs and to determine the abundance of mRNA transcripts from different tissues or under different conditions.

Preparing Equipment

Northern blotting works poorly if the mRNA is degraded to any extent. To avoid this problem, be sure all solutions are either certified RNase-free, or that solutions have been treated with DMPC (DEPC is a more conventional but also more toxic alternative.) All pipet tips should be RNase-free; electrophoresis apparatus (including the gel boxes, combs, and gel molds) and pipettors should be treated with a RNase inactivating solution (e.g., RNaseZap™ or RNase Away™). Always wear gloves when handling RNA, and change gloves frequently.

In this laboratory exercise, you will be determining the size of the *PGK* mRNA. RNA, isolated in laboratory 10, will be run on an agarose gel, transferred to a nylon membrane, then hybridized to a labeled *PGK* probe.

Precautions

Many of the reagents used in this exercise are potentially toxic. Be sure to wear gloves and take other necessary precautions when handling the chemicals in this laboratory.

Materials

- Loading buffer (RNase-free)
- RNase removal solution (e.g. RNase Away™, RNaseZap™, etc.)
- 1 X TBE buffer (made with RNase-free water; this solution should not be treated with DMPC or DEPC)
- Agarose
- Agarose gel electrophoresis apparatus
- Guanidine thiocyanate
- Nylon membrane (Ambion or Boehringer Mannheim)
- 68°C shaking water bath
- Sealable bags
- Bag sealer
- RNase-treated plastic or glass dishes for washing the membrane

- RNA size marker, preferably digoxigenin-labeled (e.g., Boehringer Mannheim Cat. #1373099)
- Gloves
- Chelex® resin
- *PGK* up and *PGK* down primers (as used in the RT-PCR lab 11)
- Digoxigenin dNTP labeling mix (as used in probe labeling, lab 6)
- *Taq* polymerase buffer, *Taq* DNA polymerase
- p20, p200 pipettors, treated to be RNase-free
- RNase-free pipet tips (for p20, p200)
- High SDS prehybridization/hybridization buffer
- Washing Solution I, II, Washin, buffer blocking solution, Western Blue® or Color Substate buffer (appendix)

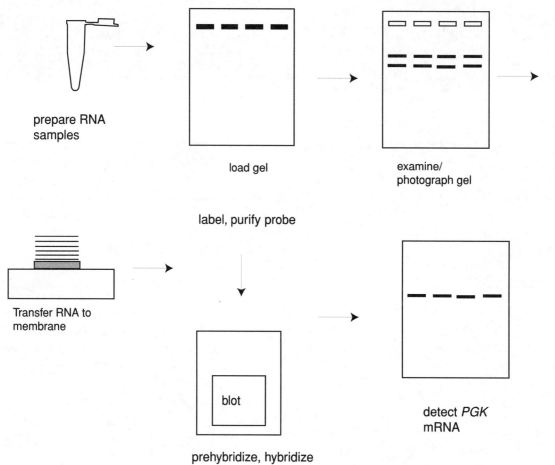

Figure 12.1 Overall procedure for Lab 12.

Procedures (See Figure 12.1)

RNA Isolation

Procedure from Lab 10; there should be sufficient RNA from lab 10 for the both the RT-PCR reaction and this lab exercise.

Probe Preparation

Make a probe as follows:

Do PCR using the *PGK* up and *PGK* down primers described in the RT-PCR lab. Grow yeast cells on Sabouraud Dextrose Agar. Pick a colony with an inoculating loop. Disperse the cells in 100 μl of TE buffer. Add 50 μl of acid-washed glass beads (425–600 microns). Vortex mix for two minutes. Centrifuge the solution for 5 min. at maximum speed. Pipet out the supernatant to a new tube. Add a small

amount (~ 5 μl in volume) of Chelex-100® resin. Boil solution for 10 min. Centrifuge for 10 min. at maximum speed. Pipet out 2 μl of supernatant, add to 0.2 or 0.5 ml PCR tube containing:

- 1 μl of *PGK* up primer (25 pmoles)
- 1 μl of *PGK* down primer (25 pmoles)
- 5 μl of dig-dNTP labeling mix (1 mM each dNTP except 0.65 mM dTTP, 0.35 mM digoxigenin-labeled dUTP)
- 5 μl of 10X PCR buffer with MgCl$_2$ (2.5 mM MgCl$_2$, final conc.)
- 38 μl sterile distilled or deionized water
- 0.5 μl of *Taq* DNA polymerase (5 units/ml).

Overlay with mineral oil if required. Put in a thermocycler for 35 cycles of 94°C for 30 Sec., 55°C for

30 Sec., 72°C for 1 min. Samples can be run overnight with the thermocycler set to hold the samples at 4°C after the run is completed. Store samples at −20°C until ready to be used.

Analyze the PCR reaction on an agarose gel to verify that a product was made, and verify labeling using a control strip as described in Laboratory 6. Assuming a product was made, purify the PCR product by electrophoresis through an agarose gel, followed by purification through a spin column (Lab 5) or by gel filtration of the PCR product (described in Lab 16, linker ligation).

Electrophoresis
1 X TBE Buffer
Add together, in a RNase-free bottle, 10.8 g Tris base, 5.5 g of Boric Acid, and 0.74 g $Na_2EDTA \cdot 2H_2O$. Add distilled or deionized, DMPC-treated water to 1 liter.

RNase-free Loading Buffer
50% glycerol, 1 mM EDTA, 0.4% Bromophenol blue, 0.4% Xylene cyanol

In an autoclavable test tube or bottle, mix together 5 ml of 100%, RNase-free glycerol, 4 ml deionized or distilled water, 20 µl 0.5 M EDTA, pH8, 40 mg Bromophenol blue, 40 mg Xylene cyanol. Add 10 µl of DMPC, bring volume to 10 ml, mix, allow to sit at room temperature for 30 minutes. Autoclave 15 min. at 15 lbs of pressure.

Alternatively, you can use a RNase-free nucleic acid loading buffer available from many biotechnology companies.

Transfer Buffer: (7.5 mM NaOH)
Per liter: Add 0.3 g NaOH to 1 liter DMPC-treated water. Mix to dissolve.

Preparation of a Gel for Northern Analysis

(1.5% Agarose in 1 X TBE, 10 mM guanidine thiocyanate.)

For a 25 ml gel add 0.38 g of agarose to 25 ml of 1X TBE buffer. Melt the agarose as in previous laboratory exercises. Then add 30 mg of guanidine thiocyanate, 2.5 µl of a 5 mg/ml ethidium bromide solution, and pour the gel.

RNA Sample Preparation

Use 20 µg of your RNA sample. Resuspend the RNA in a total volume of 10–20 µl by adding DMPC-treated sterile water. Heat at 68°C for 5 min. Spin briefly in a microcentrifuge. Place on ice, add 1 µl loading dye per 10 µl of RNA sample. Load the samples on the gel. Prepare 1 tube with RNA markers for each gel being run.

Sample Loading, Gel Running

Load the full volume of the samples on the gel (if possible) with the RNA markers either in the middle or on either side of the gel.

Run gel ~ 1.5 hours @ 3.5 V/cm (e.g., ~ 56 V for an OWL B1A apparatus)

While the gel is running, prepare the transfer apparatus as described in the Southern blotting lab exercise (lab 4). Use 7.5 mM NaOH as the transfer buffer. Be sure to use positively charged membranes (e.g., Ambion Charged Nylon membrane).

Blotting

Remove the gel from the apparatus, visualize on a transilluminator, photograph.

Then bring the gel to the transfer apparatus, gently slide the gel out of the gel mold and transfer overnight to positively charged nylon membrane. (As in Lab 4).

After transfer, wash the membrane for 2 minutes in 20X SSC. UV-crosslink the RNA to the membrane by placing the membrane on a piece of plastic food wrap on top of the transilluminator (RNA side toward the transilluminator) and irradiating for 3 minutes. The rRNA bands should be visible on the membrane. If you didn't run a marker RNA on the gel, mark the positions of the rRNA bands with a pencil.

The blot can be dried and stored at room temperature, wrapped in plastic wrap, until the next laboratory period.

Pre-hybridization, Hybridization, Washing

High SDS Prehybridization/Hybridization Buffer

(7% SDS, 5X SSC, 50 mM sodium phosphate, pH 7.0, 0.1 mM EDTA, 2% Blocking reagent (Boehringer Mannheim))

For 100 ml, add:
- 50 ml distilled or deionized water

- 25 ml 20X SSC
- 1.95 ml of 1 M 1M NaH_2PO_4
- 3.05 ml of 1M Na_2HPO_4
- 20 μl 0.5 M EDTA, pH 8

Adjust pH to 7.0 with NaOH or HCl. Add 2 g blocking reagent. Microwave to dissolve blocking reagent (do not let solution boil). Add 7g SDS, distilled or deionized water to 100 ml. Then add 0.1 ml DMPC. Mix, allow to sit at room temperature for 30 min. Autoclave. You must heat this solution in the 68°C water bath to dissolve the solids, before it is added to the blots.

- 1M NaH_2PO_4 is prepared by adding 12 g of anhydrous, monobasic sodium phosphate to a final volume of 100 ml of distilled or deionized water.
- 1M Na_2HPO_4 is prepared by adding 14.2 g of anhydrous, dibasic sodium phosphate to a final volume of 100 ml of distilled or deionized water.

(20 X SSC prepared previously — consult Appendix)
(10% SDS prepared previously — consult Appendix)

Wash Solution I

(2X SSC, 0.1% SDS used previously — Southern blotting lab — consult appendix)

Treat this solution with DMPC before use. Add DMPC, in a working chemical fume hood, at a ratio of 1:1000 (e.g., 1 ml DMPC per 1 liter of solution) shake to mix, then allow the solution to sit for 30 minutes at room temperature. Autoclave for 15 minutes at 15 lbs pressure to sterilize and inactivate the DMPC.

Wash Solution II

(0.5 X SSC, 0.1% SDS used previously — Southern blotting lab — consult appendix)

Treat this solution with DMPC before use. Add DMPC, in a working chemical fume hood, at a ratio of 1:1000 (e.g., 1 ml DMPC per 1 liter of solution) shake to mix, then allow the solution to sit for 30 minutes at room temperature. Autoclave for 15 minutes at 15 lbs pressure to sterilize and inactivate the DMPC.

1. Add 10-20 ml of Prehybridization buffer to the blot. Put in 68°C shaking water bath for 1 hour.

2. While blot is prehybridizing, set up a boiling water bath. About 15 min. before the prehybridization is complete, boil the probe (with a Lid-lock) for 5 min. Immediately place the probe on ice.

3. Remove blot from water bath, pour off Prehybridization solution. Add 10–20 μl of probe DNA solution to 10–20 ml of fresh Prehybridization solution in a previously unused, sterile plastic tube. Pour the Hybridization solution into the bag with the membrane. Seal the bag. Incubate the blot in 68°C water bath overnight.

4. The next day, wash the membrane twice in Wash Solution I for 5 min. for the first wash, 10 minutes for the second, at 68°C in a shaking water bath.

5. Then wash the membrane twice in Wash Solution II for 10 min per wash at 68°C in a shaking water bath. Pour off the solution, then continue immediately with Step 1 in the detection section, below.

Detect RNA on the Northern Blot

The Washing Buffer, Maleic Acid Buffer, Blocking Solution, and Detection Solution for detecting the mRNA are as described in the Appendix. However, after making the solutions, add DMPC, in a working chemical fume hood, at a ratio of 1:1000 (e.g., 1 ml DMPC per 1 liter of solution) shake to mix, then allow the solution to sit for 30 minutes at room temperature. Autoclave for 15 minutes at 15 lbs pressure to sterilize and inactivate the DMPC.

Procedure

All the following steps are done at room temperature, with gentle shaking, except Step 6.

1. Place the membrane in 50 ml Washing buffer for 1 min.

2. Block by gently shaking the membrane for 30 min. in Blocking solution (15 ml per 7 X 8 cm membrane). Pour off Blocking solution.

3. Add Antibody solution (blocking buffer with 1:5000 dilution of antibody, 15 ml with 3 μl antibody per 7 X 8 cm membrane). Incubate the membrane for 30 min.

4. After 30 min, discard the Antibody solution, add 50 μl Washing buffer. Wash twice, 10 min. per wash.

5. Pour off Washing buffer. Incubate in 20 ml of detection buffer for 2 minutes. Pour off Detection buffer.

6. Add sufficient Western Blue® to cover the membrane. Place the blot in the dark, and incubate without shaking. Once the bands are sufficiently dark, pour off the Western Blue®, wash with water.

7. Using semi-log paper, plot the size of the marker RNAs (as you did in lab exercise 2). Determine the size of the PGK mRNA based on the marker DNA sizes.

Tips

New plastic or glass containers that will only be used for RNA analysis should be used for the washing steps. For glass containers, DMPC-treat. For plastic containers, rinse with RNaseZap™ or other RNase removal solution, rinse with DEPC-treated water.

Once you begin the procedure, do not allow the membrane to dry.

References

Goda, S., and N. Minton. 1995. A simple procedure for gel electrophoresis and Northern blotting of RNA. Nucleic Acids Research 23(16): 3357-3358.

Sambrook, J. E. Fritsch, and T. Maniatis. 1989. Molecular Cloning. A Laboratory Manual. 2nd Ed. Cold Spring Harbor Laboratory Press; Cold Spring Harbor, NY.

Northern Blotting

Name:_____ **Date** _____

Plot the migration versus size (in nucleotides) of the marker RNAs on the semi-log paper, then determine the size of the *PGK* mRNA, based on its migration.

Results

Questions

1. Why is the RNA denatured when run through agarose gels, but DNA is typically not denatured while it is run?

2. What advantages do you see for RT-PCR compared to northern hybridization? What disadvantages do you see?

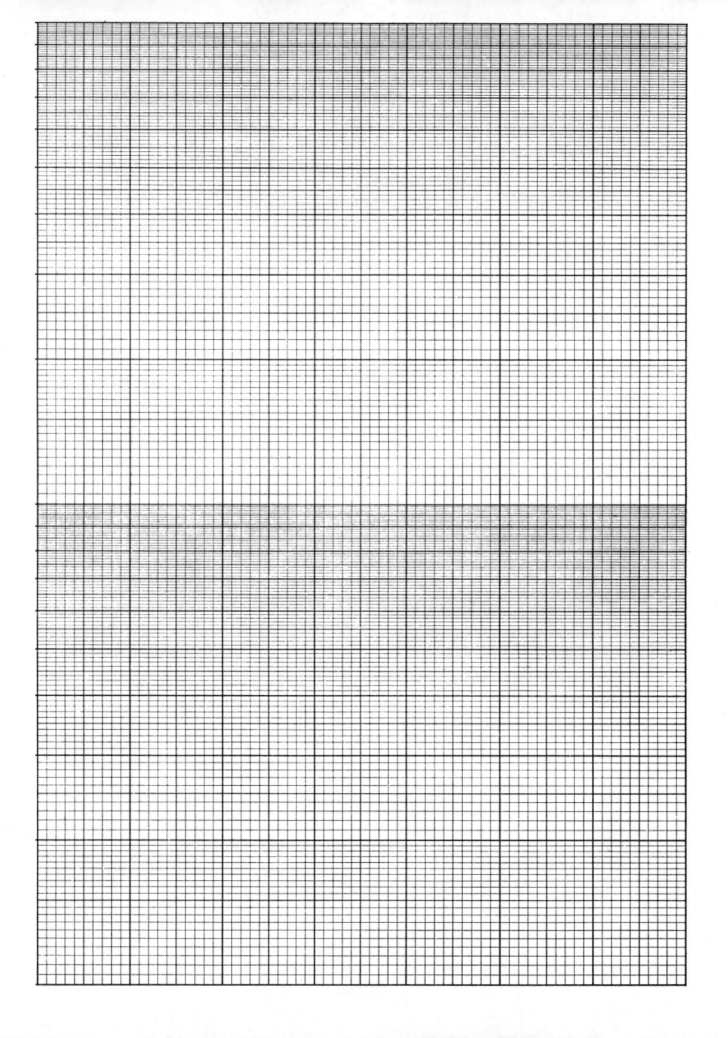

I V
DNA Manipulation & Cloning

The exercises in this section of the lab manual are designed to teach you methods for preparing and altering DNA. The laboratory exercises are focused around the plasmid pGEM®-*luc* (Promega Biotech) which has a 1.7 kb insert of the firefly (*Photinus pyralis*) luciferase gene into the plasmid pGEM®-11zf(-). The ultimate goal of this set of exercises is to alter pGEM®-*luc* so the firefly luciferase gene is efficiently expressed in *E. coli*.

The first exercises (Laboratories 13 and 14) teach you techniques for producing and isolating plasmid DNA. Laboratory 13 presents two methods for preparing *E. coli* cells that are capable of taking up plasmid DNA. Laboratory 14 then shows you two methods of isolating plasmid DNA from the cells prepared in the previous exercise.

The next set of laboratories contain a set of procedures from modifying the pGEM®-*luc* plasmid by the insertion of a linker (a short double-stranded DNA). Initially (Lab 15) you digest the plasmid and prepare the linker. Next (Lab 16) you join the linker and the plasmid together. In the final exercise (Lab 17), you will use two techniques to verify that you have altered the plasmid.

Competent Cell Preparation & Transformation

Laboratory 13

Key Terms

Competent cells: Cells that have been treated to maximize their ability to take up DNA.

Transformation: Introduction of DNA into competent bacterial cells.

Many molecular biology procedures require propagation of DNA in *E. coli*. In order for DNA to be introduced into *E. coli*, the bacteria must be made competent. This is done by cultivating the organisms to an exponential phase of growth, then harvesting cells and either treating them with calcium chloride or a high-voltage electrical pulse. These cells can then take up plasmid DNA and make many copies of the plasmid. In this exercise, you will prepare competent cells that are able to take up DNA. At the end of the exercise, you will test the competence (i.e. the transformability) of the cells you produced. The competence of cells is determined by incubating them with

Materials for Option A

- Calcium chloride solution, ice cold (sterile)
- Competent cell freezing buffer (sterile)
- Centrifuge (set to 0°C)
- 15 ml Falcon tubes, or sterile Oak Ridge tubes (sterile)
- Appropriately grown culture
- *Escherichia coli* JM109 or other suitable strains (e.g., DH5α (Life Technologies), XL-1 Blue (Stratagene, La Jolla, CA))
- Microcentrifuge tubes (sterile)
- Ice bath
- −70°C or −80°C freezer
- pGEM®-luc and pGEM®-11zf (mixed together, final concentration 2 ng/µl each plasmid)

Photographic Atlas Reference
Chapter 7

plasmid DNA, then plating the cells on a medium containing an antibiotic. If the cells are competent, they will take up the plasmid DNA (which contains an antibiotic resistance gene) and grow on media containing the antibiotic. The more bacterial colonies that grow after cells are incubated with plasmid DNA, the more competent are the cells. In addition to media containing antibiotics, it is often useful to use a medium that allows one to distinguish cells that contain an insert DNA from those that do not contain an insert DNA. In this exercise, MacConkey agar is used for that purpose. Plasmids without an insert produce red colonies in transformed bacteria; plasmids with an insert produce white colonies in transformed bacteria.

Precautions

Follow standard precautions for handling chemicals and bacterial cultures. Be sure that all materials contaminated with bacteria are disposed of in an appropriate manner, as described by your instructor.

Preparation of Solutions

Calcium Chloride Solution (10 mM Tris, pH 8, 50mm CaCl$_2$) To make 500 ml, add 3.7 g CaCl$_2$.2H$_2$O (Sigma Cat # C3306) and 5 ml, 1 M Tris, pH8 to a final volume of 500 ml in distilled or deionized water. Autoclave for 15 minutes at 15 lbs. pressure.

Competent Cell Freezing Buffer (20% glycerol (final concentration) in Calcium chloride solution)

Make 1.5 ml aliquots in sterile microcentrifuge tubes, containing 1.2 ml Calcium chloride solution (above) and 0.3 ml sterile glycerol.

79

Ampicillin Stock Solution (1000X, 150 mg/ml) Add 1.5 g of ampicillin to 10 ml of distilled or deionized water. Filter sterilize through a 0.2 μM filter. Put 1 ml aliquots into sterile microcentrifuge tubes, freeze at −20°C.

LB Broth Premixed dry powders are available from several manufacturers.

If you wish to make your own medium, to 425 ml of distilled or deionized water, add 5 g tryptone, 2.5 g yeast extract, 5 g NaCl. Mix until dissolved. Adjust pH to 7 with NaOH, adjust volume to 500 ml, autoclave.

LB Agar Use the recipe for LB broth, but add 7.5 g agar/ 500 ml. After autoclaving, allow the medium to cool to about 55–60°C before pouring plates. Pour about 25 ml into a standard size petri dish (100 X 15 mm); 500 ml should make 20 plates.

LB Broth or Agar with Ampicillin After autoclaving add 1 ml of Ampicillin (150 mg/ml) per liter of medium (e.g., 0.5 ml of ampicillin per 500 ml of medium). Be sure the medium has cooled (55° to 65°C before adding the Ampicillin. Ampicillin will deteriorate if stored in medium. Use any medium containing antibiotics within one to two weeks, and store at 4°C prior to use.

MacConkey Agar with Ampicillin Add the appropriate amount of medium as directed by manufacturer. Before pouring plates, add ampicillin as directed above. Be sure to use standard MacConkey agar (i.e., containing crystal violet).

Procedures

Option A

Standard method of preparing competent cells (see Figure 13.1). 0.5 to 1 ml of an overnight *E. coli* JM109 culture was inoculated into 50 ml of LB about 2–4 hours before the start of the laboratory period. This was done because *E. coli* cells are most transformable when they are growing exponentially.

1. Remove 1 ml of *E. coli* culture from the flask to determine the optical density (O.D.) at 660 nm. Do this aseptically, but without removing the culture from the 37°C incubator. The O.D. should be 0.3 to 0.45. If it is lower than that, allow cells to grow for an additional period. Note that the cells double every 20 minutes under these conditions and therefore the O.D. will also approximately double every 20 minutes.

2. When the cells are at the appropriate O.D., pour 15 mls of culture into a sterile, 15 ml Falcon® tube, and chill the suspension on ice for 10 min. (Three students or groups can share the same flask.) From this point until the cells are transformed, they must be kept COLD. If the cells are allowed to warm above ice temperature, the competence of the cells will decrease.

3. While waiting for the cells to cool, place two LB Agar with ampicillin plates and two MacConkey with ampicillin plates in a 37°C incubator.

4. Centrifuge your 15 ml culture of cells for 10 min. at 4000 rpm. Pour off supernatant into an appropriate biohazard waste container.

5. Resuspend the bacterial pellet by flicking the bottom of the tube. The cells must be resuspended completely. Add 15 ml of an ice-cold calcium chloride solution. Place the cell solution on ice for 10 min. Centrifuge for 10 min. at 4000 rpm.

6. Pour off supernatant, flick the bottom of the tube to resuspend. *Do not vortex mix.* Add 0.5 ml of Competent Cell Freezing Buffer. Aliquot 100 μl to each of your pre-chilled microcentrifuge tubes (5 or 6 tubes total), marked with your initials and the date. Leave two tubes on ice. Put the remainder at −80°C. Then follow the DNA transformation procedure, below.

Option B

Simplified method of producing competent cells (See Figure 13.1B)

Materials

- Sterile plastic or metal inoculating loops
- Sterile microcentrifuge tubes
- Sterile CaCl$_2$ solution (prepared as described in the previous section)
- *Escherichia coli* JM109 or other suitable strains (e.g., DH5α (Life Technologies) or XL-1 Blue (Stratagene, La Jolla, CA))
- Ice bath
- pGEM®-*luc* and pGEM®-11zf(-) (mixed together, final concentration of 2 ng/μl for each plasmid)

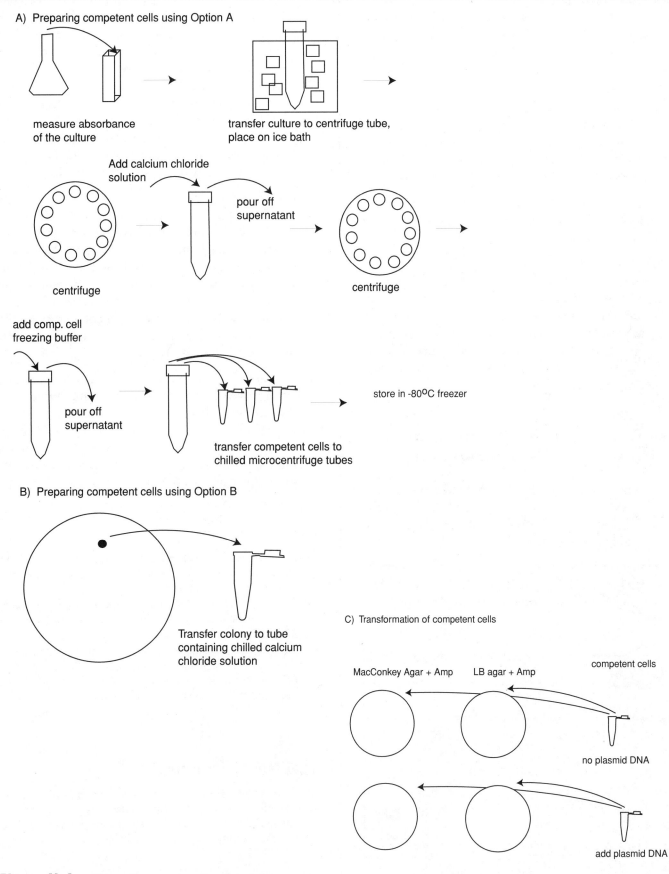

A) Preparing competent cells using Option A

measure absorbance
of the culture

transfer culture to centrifuge tube,
place on ice bath

centrifuge

Add calcium chloride
solution

pour off
supernatant

centrifuge

add comp. cell
freezing buffer

pour off
supernatant

transfer competent cells to
chilled microcentrifuge tubes

store in -80°C freezer

B) Preparing competent cells using Option B

Transfer colony to tube
containing chilled calcium
chloride solution

C) Transformation of competent cells

MacConkey Agar + Amp   LB agar + Amp   competent cells

no plasmid DNA

add plasmid DNA

Figure 13.1   Overall procedure for Lab 13.

Procedure for Option B

Grow *E. coli* JM109 (or other appropriate strain) overnight (at least 20 hours) at 37°C on LB agar without antibiotics. After overnight growth, the plates can be refrigerated for at least several days before the subsequent procedure, if necessary.

At least 30 minutes to an hour in advance of the transformation, put two LB Agar with ampicillin plates and two MacConkey agar with ampicillin plates in a 37°C incubator. The plates must be at 37°C before you start the transformation procedure.

1. Transfer 200 µl of ice-cold CaCl₂ solution to each of two sterile microcentrifuge tubes. Keep the tube on ice.

2. Transfer a large colony from the plate to each sterile microcentrifuge tube. Mix the cells vigorously by twirling the loop rapidly in the CaCl₂ solution. If any cell clumps are present, break them up by gently pipetting up and down with a micropipet. The solution should now be cloudy. Keep the tube on ice. Then follow the transformation procedure, below.

DNA Transformation: (for cells prepared by either option A or B, Figure 13.1C)

1. To one of the tubes of competent cells on ice, add nothing. To the other tube, add 5 µl of the pGEM®-*luc*, pGEM-11zf(-) solution (10 ng each plasmid, 20 ng of plasmid DNA total).

2. Plate 50 µl (option A) or 100 µl of cells (option B) from the "no DNA" tube to a pre-warmed MacConkey agar plate with ampicillin. Then plate 50

µl (Option A) or 100 µl (Option B) of the cells from the same tube to pre-warmed LB agar with ampicillin. *It is critical that the plate is at 37°C when you inoculate with your cells.* Spread with a sterile spreader or inoculating loop to evenly distribute bacteria over the surface of the plate. Label the plates with your initials and "-" DNA. Plate cells from the tube containing pGEM®-*luc* and pGEM-11zf(-) DNA to separate MacConkey and LB agar plates as above. Put your initials and "+ DNA" on the bottom or side of the plates.

3. Return the plates immediately to the 37°C incubator. The next morning, take plates out (cells should not be allowed to grow for more than 20 hours) and mark the red colonies on the bottom of the MacConkey agar plates. Count the colonies on the LB ampicillin plate. Refrigerate the plates. The day before the next laboratory exercise, inoculate two red colonies and two white colonies from the MacConkey plate to LB broth containing 150 µg/ml ampicillin (see Lab 14).

4. Calculate the number of transformants per µg of DNA (based on the number of colonies present on the LB amp plate). Remember that a ng is 10^{-9} grams; a µg is 10^{-6} grams. Commercially-available competent cells have transformation efficiencies ranging from 10^6–10^{10} per µg of DNA. Your cells should give at least 5×10^3 transformants per µg.

References

Jennings, M. and I. Beacham. 1989. MacConkey Agar as an alternative to X-gal in the detection of recombinant plasmids. Biotechniques 7:1082.

Pope, B. and H. Kent. 1996. High efficiency 5 min. transformation of *Escherichia coli*. Nucleic Acids Research. 24(3): 536-537.

Taylor, K-A. 1996. Comparison of a five-minute *E. coli* transformation protocol with the conventional method of Mandel and Higa. MBI Fermentas Nucleic Type Newsletter, No. 1. MBI Fermentas, Inc., Amherst, NY.

Theoretical Treatment

Hanahan, D. and F. Bloom. 1996. Mechanisms of DNA transformation. pps2449-2459 in Neidhardt, F., ed. *Escherichia coli and Salmonella*. Cellular and Molecular Biology. Volume 2. Washington, D. C.: ASM Press.

Materials

For transformation

- 37° incubator
- pGEM-*luc* and pGEM-11zf (−), mixed together (final concentration of 2 ng/µl for each plasmid)
- LB Agar
- Ampicillin
- MacConkey Agar
- Sterile plastic petri plates

Competent Cell Preparation and Tranformation

Name:_____   **Date** _____

Results

_____ Transformants/μg of DNA (Show calculations.)

Questions

1. Provide two possible explanations if your "no DNA" competent cell test produced colonies on plates.

2. Most *E. coli* strains that are used to make competent cells contain a mutation in an endonuclease. Why would this mutation be useful?

Plasmid DNA Isolation

Plasmid: A circular DNA molecule used for transferring foreign DNA into bacterial cells.

Plasmid preparation: A procedure for isolating purified plasmid DNA from bacterial cells.

In this lab, you will prepare plasmid DNA from the cells you had transformed in the previous laboratory exercise. You will do two things with this DNA; 1) look at differences in the restriction enzyme digestion pattern of the pGEM®-*luc* and pGEM®-11zf(-) DNAs and 2) use the pGEM®-*luc* DNA in a subsequent laboratory exercise.

Precautions

- This procedure requires the use of NaOH solution, a strong base. Wear appropriate protective clothing as directed by your instructor.
- Dispose of all bacterially-contaminated cultures as directed by your instructor.

Materials for Option A

- Tubes of competent cells from previous laboratory period (or a plate with colonies)
- TSE (Tris, Sucrose, EDTA; sterile)
- Alkaline SDS
- 7.5 M Ammonium acetate
- RNase mixture (RNase1 and RNaseA — commercial preparations are available)
- Isopropanol
- *Eco*RI restriction enzyme and buffer
- Sterile, distilled water
- TE buffer
- Microcentrifuge tubes, sterile
- p20, p200, p1000 pipettors and sterile tips

Photographic Atlas Reference
Chapter 8

- Follow the precautions for agarose gel electrophoresis as described in Laboratory 2.

Procedures (See Figure 14.1)

The *day before this exercise*, inoculate 2 white colonies and 2 red colonies from your MacConkey Agar plate (Lab 13) into separate tubes containing 5 ml of LB broth with 150 μg/ml ampicillin. (You are inoculating two tubes just in case one doesn't grow up.) Incubate in a shaking incubator overnight at 37°C. Save the plates; a colony from the MacConkey plate will be used in Lab 17.

Option A: Laboratory-Prepared Solutions
Preparation of Solutions

TSE *(Stock solutions)*

3 M Sucrose: For a 100 ml stock solution, add 102.7 g of Sucrose to an autoclavable glass container. Add distilled or deionized water to 100 ml. Autoclave.

1 M Tris: prepared previously (see Appendix, preparation of TE buffer)

0.5 M EDTA: pH 8, prepared previously (see Appendix, preparation of TE buffer)

TSE solution: 0.3M sucrose, 25 mM Tris pH8, 25 mM EDTA, pH8

For 400 ml add:

- 330 ml distilled or deionized water
- 10 ml 1 M Tris, pH8
- 20 ml 0.5 M EDTA, pH8
- 40 ml 3 M Sucrose.

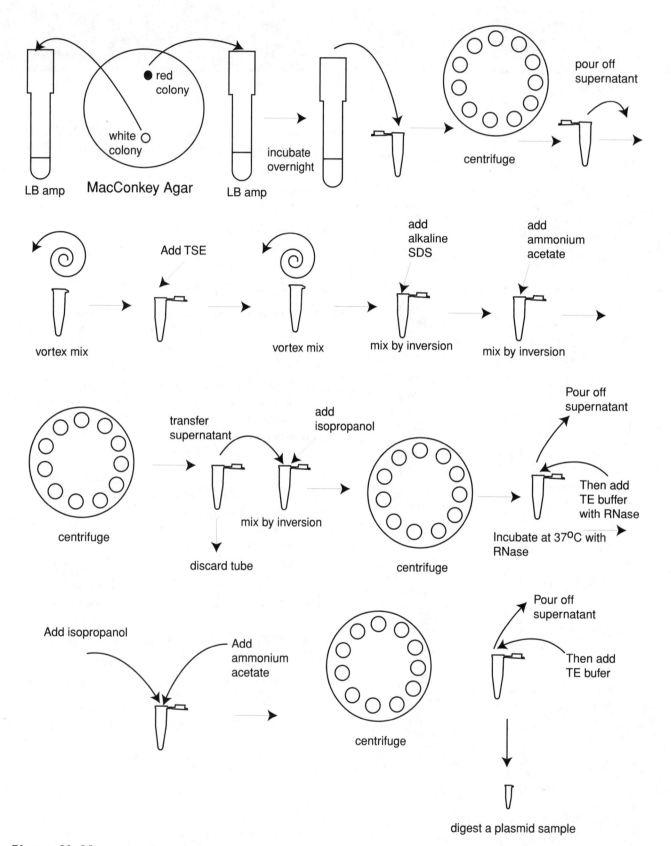

Figure 14.1A Overall procedure for Lab 14. Laboratory prepared solutions.

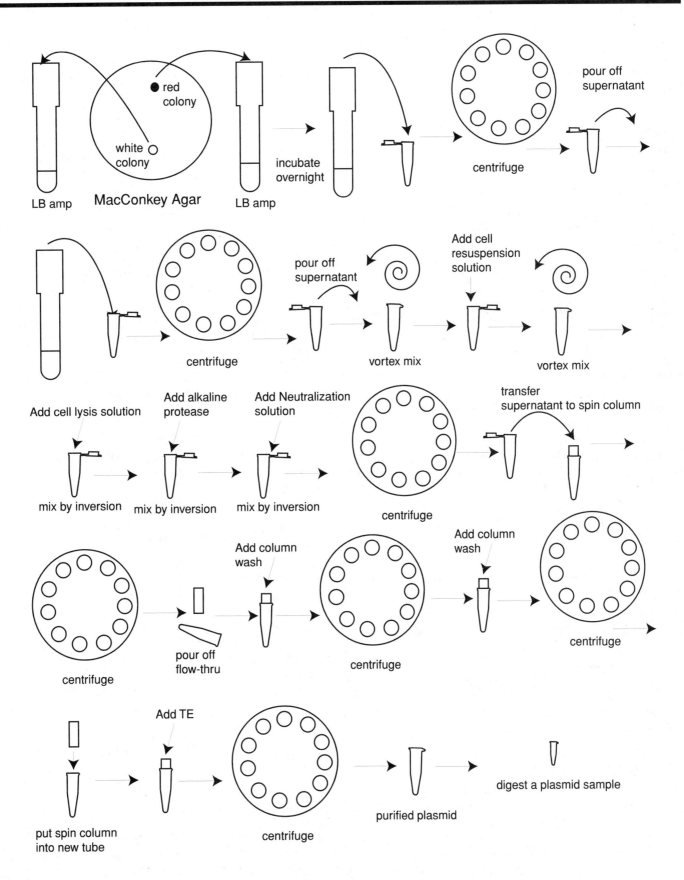

Figure 14.1B   Overall procedure for Lab 14. Plasmid preparation kit.

Alkaline SDS: 0.2 M NaOH, 1% SDS

Stock Solutions

1 M NaOH: For 100 ml, add 4 g of NaOH pellets to 90 ml distilled or deionized water. Allow pellets to dissolve, add water to bring volume to 100 ml.

10% SDS — prepared previously (see Lab 7)

Alkaline SDS: Working solution (prepared immediately before use)

Mix together, in order

- 1.4 ml distilled or deionized water
- 200 μl 10% SDS
- 400 μl 1M NaOH

This is sufficient solution for 5 plasmid preparations.

7.5 M Ammonium acetate— prepared previously (see Lab 5)

Procedures

1. Choose one test tube with cells from a white colony, one test tube with cells from a red colony. The directions that follow are for a single plasmid preparation. Simply duplicate the procedure for your second tube. Pour 1.5 ml of the culture into a sterile 1.7 ml microcentrifuge tube. Place your tube in the microcentrifuge. *Be sure the centrifuge is always balanced with an equal number of tubes on either side of the rotor before the centrifuge is started. Use a tube filled with water to balance the centrifuge, if necessary.* Centrifuge at maximum speed (12,000 X g, ~13-14,000 rpm) for 2 minutes. Pour off supernatant into an appropriate biohazard waste container. Thoroughly Vortex pellets to resuspend.

2. Add 200 μl TSE. Make sure pellet is well dispersed by vortex mixing or your yield of plasmid will be very low. Allow to sit at room temperature for 5 min.

3. Add 400 μl Alkaline SDS. Mix by inversion. **DO NOT VORTEX MIX!!** The suspension should clear. Leave no more than 5 minutes at room temperature between the addition of the alkaline SDS and step 4.

4. Add 300 μl 7.5 M Ammonium acetate, mix by inversion, chill 5 min. on ice, microcentrifuge 10 min. The plasmid DNA is in the supernatant.

5. Pour the supernatant to a new tube (discard the old tube with the white pellet). Add 700 μl isopropanol to the supernatant, put on ice 10 min., spin 10 min. The pellet now contains the plasmid DNA. Dry pellet by inversion on Kimwipes (about 5 min.).

6. Add 50 μl of TE (with 1 μl RNaseA and RNase T1 mixture) to pellet. Incubate for 15 min. at 37°C. Then add 150 μl TE, 100 μl Ammonium acetate, and 300 μl isopropanol, put on ice 10 min., then spin 10 min. The plasmid is again in the pellet. Pour off supernatant, dry on Kimwipes. Resuspend in 50 μl TE.

7. Add 10 μl of purified DNA to a sterile 0.5 ml or 1.7 ml microcentrifuge tube (one tube for pGEM®-11zf(−), and a separate tube for pGEM®-*luc*). To each tube, add 1.2 μl of 10X *Eco*RI restriction enzyme buffer, 1 μl of *Eco*RI restriction enzyme. Allow the DNAs to digest at 37°C for at least one hour or overnight, preferably in an air incubator (not a water bath). After the digestion is completed, store the samples at −20°C until they can be analyzed by gel electrophoresis.

Then run the samples on a 0.7% agarose gel containing ethidium bromide as described in Laboratory 2. Each person will need two lanes for running their samples.

Option B: Plasmid Miniprep Kit

The description given below is for the Promega Wizard® SV Miniprep kit. Follow the manufacturers' directions if you are using another kit.

Procedure

Based on Wizard® SV Minipreps DNA Purification System Protocol):

If using the kit for the first time, add 35 ml of 95% ethanol to the Column Wash Solution bottle.

1. Choose one test tube with cells from a white colony, one test tube with cells from a red colony.

Materials for Option B

- Kit components, plus 35 ml of 95% ethanol for the 50 prep system
- TE buffer

The directions that follow are for a single plasmid preparation. Simply duplicate the procedure for your second tube. Pour 1.5 ml of the culture into a sterile 1.7 ml microcentrifuge tube. Place the tube in the microcentrifuge. *Be sure the centrifuge is always balanced with an equal number of tubes on either side of the rotor before the centrifuge is started. Use a tube filled with water to balance the centrifuge, if necessary.* Centrifuge at maximum speed (12,000 X g, ~13-14,000 rpm) for 2 minutes. Pour off the supernatant into a designated biohazard waste container. Pour in another 1.5 ml of the culture into the same microcentrifuge tube. Centrifuge as described above, pour off the supernatant. At this point you should have a cell pellet in the bottom of your tube. This pellet should represent cells from a total of 3 ml of culture.

2. Vortex mix cells thoroughly to completely resuspend the cell pellet. This will probably take 30 seconds to 1 minute for each tube. Add 250 µl of Cell Resuspension solution. Vortex mix again until no cell clumps are visible, and the pellet has been completely removed from the bottom of the centrifuge tube.

3. Add 250 µl of Cell Lysis solution. Mix by inversion 4 times. **DO NOT VORTEX MIX** this solution, or any solution in the following steps.

4. Add 10 µl of Alkaline Protease solution and mix by inverting the tube 4 times. Incubate 5 minutes at room temperature.

5. Add 350 µl of Neutralization solution. Mix by inversion 4 times.

6. Centrifuge the cell lysate at maximum speed in a microcentrifuge for 10 minutes at room temperature. The plasmid DNA is in the supernatant.

7. Pour the supernatant into the spin column. (Discard the tube containing the white pellet.) Insert the spin column into an open-top 2 ml centrifuge tube (collection tube).

8. Centrifuge the supernatant at maximum speed for 1 minute at room temperature. The plasmid DNA is bound to the resin in the spin column. Discard the liquid that was centrifuged into the collection tube. Place the spin column back into the collection tube.

9. Wash the plasmid DNA. Add 750 µl of Column Wash solution (containing ethanol) to the spin column, centrifuge at maximum speed for 1 min. at room temperature. (The DNA remains bound to the resin in the spin column.) Discard the liquid in the collection tube. Place the spin column back into the collection tube.

10. Wash the DNA a second time. Add 250 µl Column Wash solution to the spin column, centrifuge at maximum speed for 2 minutes. (The DNA remains bound to the resin in the spin column.) Discard the collection tube (and liquid), place the spin column in a new, sterile 1.5 ml microcentrifuge tube. You may wish to cut off the cap of the tube, prior to inserting the spin column. In some centrifuges, the open lids catch and break, sometimes pulling the tubes out with them.

11. Add 100 µl of TE (or nuclease-free water) to the spin column to elute the plasmid DNA. Centrifuge the tube at room temperature for 1 minute at maximum speed. The purified plasmid DNA should be in the liquid in the bottom of the tube. The purified DNA should be stored at −20°C (except for the aliquot used in the next step).

12. Add 5 µl of purified DNA solution to a sterile 0.5 ml or 1.7 ml microcentrifuge tube (one tube each for pGEM®-11zf(-), one tube for pGEM®-*luc*). To each tube, add 3 µl of distilled water, 1 µl of 10X *Eco*RI restriction enzyme buffer, 1 µl of *Eco*RI restriction enzyme. Allow the DNAs to digest at 37°C for at least one hour or overnight, preferably in an air incubator (not a water bath). After the digestion is completed, store the samples at −20°C until they can be analyzed by gel electrophoresis. Then run the samples on a 0.7% agarose gel containing ethidium bromide as described in Laboratory 2. Each person will need two lanes for running their samples.

References

Guilfoile, P. and S. Plum. In Press. Using a genetic selection and genetic screen to illustrate the relationship between phenotype and genotype: A DNA transformation and DNA isolation laboratory exercise. The American Biology Teacher.

Lee, S. and S. Rasheed. 1990. A simple procedure for maximum yield of high-quality plasmid DNA. Biotechniques 9: 676-679.

Promega Biotech. 1996. Wizard® Plus SV Miniprep DNA purification system protocol. Madison, WI.

Plasmid DNA Isolation

Laboratory 14

Name:_____ **Date** _____

Results

Questions

1. What is different about the properties of proteins and nucleic acids that allows them to be separated?

2. Based on this laboratory exercise, what do you know about the relative solubility of nucleic acids in water and alcohol?

3. The addition of alkali is essential for separation of genomic DNA and plasmid DNA. Explain why this is so.

4. Most plasmid isolation kits use some type of resin or matrix for binding plasmid DNA. List and describe two likely properties of these resins or matrices.

Linker-based Mutagenesis

15

Key Terms

mutagenesis: a method for introducing DNA sequence changes. In this laboratory exercise, mutagenesis involves deleting a piece of DNA from the pGEM®-*luc* plasmid, followed by the insertion of a different piece of DNA.
linker: a short, synthetic, double-stranded section of DNA, used to join or bridge two DNA segments.

E. coli cells containing the pGEM®-*luc* plasmid do not efficiently express the luciferase enzyme because there is no prokaryotic ribosome binding site upstream of the luciferase (*luc*) gene, and the upstream *lacZ* gene is in a different reading frame than the *luc* gene. The goal in this exercise is to genetically engineer the pGEM®-*luc* plasmid so it is efficiently expressed in *E. coli*. This will be accomplished by inserting a synthetic piece of DNA into the pGEM®-*luc* plasmid that will put the luciferase gene into the same reading frame as the *lacZ* gene. This can be detected by the new *Xho*I site in the linker, and by the ability of *E. coli* cells containing the engineered plasmid to glow due to luciferase expression.

Figure 15.1 shows that, in the process of generating this plasmid, you will delete several restriction enzyme sites, including *Bam*HI, and you will add an

Materials

- pGEM®-*luc* plasmid (prepared in the previous laboratory exercise, or from Promega Biotech)
- Linker A: 5'-AGCTTAACTCGAGA-3'. (dissolved in TE at a concentration of 20ng/μl)
- Linker B: 5'-GATCTCTCGAGTTA-3' (dissolved in TE at a concentration of 20ng/μl)
- *Bam*HI restriction enzyme and buffer
- *Hin*dIII restriction enzyme and buffer
- pipet tips for p20, p200 pipettors
- LB agar, ampicillin stock solution

Photographic Atlas Reference
Chapter 8

*Xho*I site. These changes will aid in the construction and analysis of the modified plasmid. Linearized DNA transforms poorly. Therefore, if you digest the ligation mix with *Bam*HI, plasmids without the linkers will be linearized, whereas plasmids with the linkers will not be linearized. The only efficiently transforming DNA should then be the circular plasmid containing the linkers. You can readily distinguish this modified plasmid from the original plasmid by digestion with the restriction enzyme *Xho*I.

Precautions

Follow standard precautions for handling laboratory chemicals.

Procedure

1. Based on the gel analysis from the previous lab, pick DNA samples from the class that yielded substantial amounts of pure DNA from the pGEM®-luc plasmid preparation. Mix together:

   - 16 μl of pGEM®-*luc*
   - 2 μl *Bam*H1 buffer
   - 1 μl *Bam*H1
   - 1 μl *Hin*dIII.

   Digest the plasmid overnight or for at least one hour at 37°C.

2. Linker annealing. During this step, the two single-stranded DNAs will join together to form a double-stranded DNA molecule. This double-stranded DNA will have complementary ends to the *Hin*dIII, *Bam*HI digested pGEM®-*luc* plasmid.

After the digest has begun, add 5 μl of linker A and 5 μl of linker B to a 0.5 ml microcentrifuge tube. Heat to 70°C for 10 min. Then put the tube in a beaker of 70°C water, and allow the linker to slowly cool to room temperature. Mark the tube "ds linker", freeze at −20°C.

3. Pour LB plates with 150 μg/ml ampicillin (as described in Lab 13).

Reference

Promega Biotech Catalog. Madison, WI.

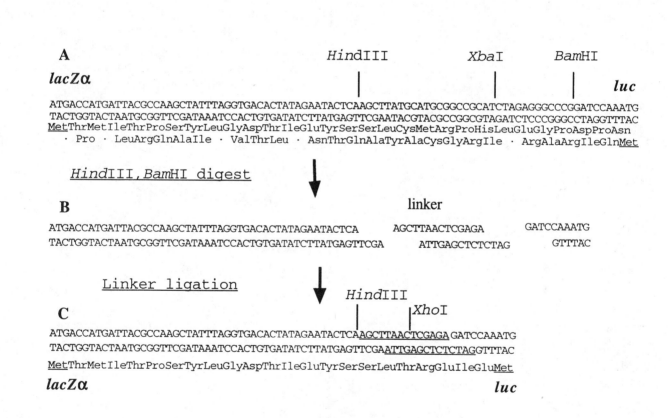

Figure 15.1 Overall procedure of laboratory 15. Panel A shows the sequence of a portion of the original pGEM®-*luc* plasmid. The underlined Methionine codons represent the start of the *lacZ* gene and the *luc* gene respectively. Note that these two genes are in different reading frames (the dots in the "*luc*" reading frame represent stop codons). Panel B shows the sequences involved in the ligation reaction. Panel C shows the sequence of the altered plasmid. Note that in Panel C, both the *lacZα* and *luc* genes are in the same reading frame. The underlined sequence is derived from the linker.

Key Term

DNA ligation: Covalent joining of DNA molecules, catalyzed by the enzyme DNA ligase.

In this exercise, you will be joining (ligating) the digested pGEM®-*luc* plasmid to the linker you prepared in the previous laboratory exercise. If the ligation reaction works properly, the modified plasmid should now allow *E. coli* to efficiently express the luciferase protein. Luciferase expression will be tested in the next laboratory exercise.

Procedures (See Figure 16.1)

1. Heat the digest of pGEM®-*luc* to 65°C for 10 min. This inactivates the *Hin*dIII restriction enzyme. If the ligation works, there will be no *Bam*HI site present in the plasmid, so the *Bam*HI doesn't need to be inactivated (*Bam*HI requires heating to 80°C for inactivation).

2. Purify the pGEM®-*luc* DNA from the small DNA fragment released by restriction enzyme digestion.

Materials

- 65°C water bath
- ds linker DNA (prepared in the previous lab exercise)
- Digested pGEM®-*luc* (prepared in the previous lab exercise)
- Centri Spin 40 columns (Princeton Separations, Adelphia, N.J. Cat. #CS-401)
- 0.5 ml microcentrifuge tube (for ligation reaction)
- T4 DNA ligase, ligase buffer
- *Bam*HI restriction enzyme, restriction enzyme buffer
- LB ampicillin plates (prepared in the previous lab exercise)

Photographic Atlas Reference
Chapter 9

This is done by purification through a spin column. For this procedure, add 500 µl of TE to the spin column, allow the gel to hydrate for 30 min. Flick the tube to remove bubbles. Then take off the top and bottom caps, orient the column so the ridge points up, spin at 750 X g (3000 rpm in a Eppendorf® 5415C centrifuge) for 2 min. KEEP THE COLUMN, discard the tube containing the liquid. Put the column in a new microcentrifuge tube, load restriction enzyme digest (20 µl) directly to the middle of the gel. Spin again at 750 X g for 2 min. KEEP THE LIQUID flow through (purified pGEM®-*luc*), discard the column. This procedure gets rid of the small section of the pGEM®-*luc* plasmid that was released by *Bam*HI, *Hin*dIII digestion. This section of DNA could be re-ligated into the plasmid if it was not removed.

3. Mix together:
   - 5 µl of the linkers (half of the reaction prepared in Laboratory15)
   - 12 µl pGEM®-*luc* (purified, digested with *Hin*dIII and *Bam*HI)
   - 2 µl 10 X ligase buffer
   - 1 µl T4 DNA ligase

4. Allow the reaction to sit for at least 1 hour or overnight at room temperature.

5. After the ligation reaction is complete, add 2 µl *Bam*HI restriction enzyme buffer, 1 µl *Bam*HI enzyme. Incubate 30 min. to one hour (or overnight) at 37°C. Transform competent cells (prepared in Lab 13) with the ligation mix, or freeze the ligation mix at −20°C until you are ready to do the transformation.

6. Thaw frozen competent cells on ice. Do the transformation as directed in Laboratory 13, plating

the cells on LB medium containing 150 µg/ml of ampicillin. Invert and incubate plates at 37°C overnight; the plates can be refrigerated after overnight incubation. The day *before* the next lab period, you will need to inoculate 2–5 different colonies into test tubes containing sterile LB broth with ampicillin.

References

New England Biolabs Catalog. Beverly, MA.

Princeton Separations. Protocol for use of Spin columns. Adelphia, N.J.

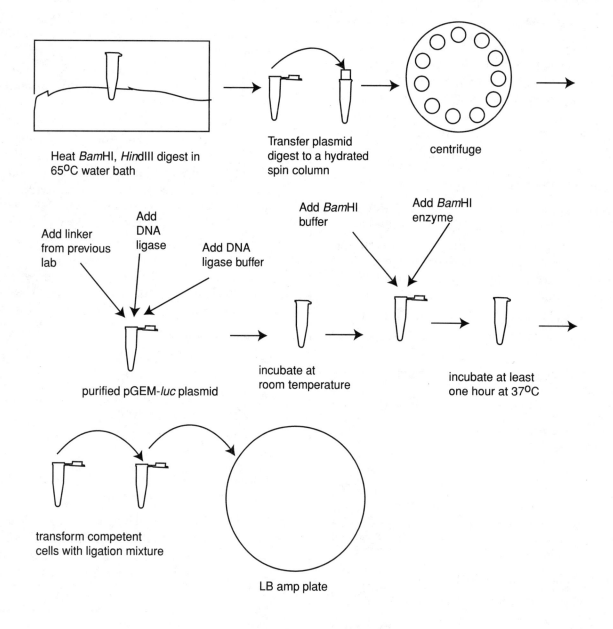

Figure 16.1 Overall procedure for Lab 16.

pGEM®-luc, Linker DNA Ligation

Name:_____   Date _____

Results

Questions

1. Would this procedure be successful if you used oligonucleotides which each contained an extra base, compared to the oligonucleotide used in the laboratory exercise? Why or why not?

2. In the lab, the digestion was done with *Bam*HI in the ligation reaction. Would this procedure have worked if you had substituted *Hin*dIII or *Xho*I enzymes for *Bam*HI? Why or why not?

3. Why were transformants plated to LB medium, rather than LB medium with X-gal or MacConkey agar?

Luciferase is an enzyme naturally produced by the firefly, *Photinus pyralis*. In the presence of its substrate (luciferin), this enzyme catalyzes production of light. In addition to its use by the firefly, luciferase is also widely used as a reporter gene — a gene which can be coupled to a regulatory DNA sequence to determine how the regulatory sequence functions.

In this laboratory, you will test whether you have successfully modified the pGEM®-*luc* plasmid. The first test will be a functional test—to see if bacteria that contain the modified plasmid glow when the luciferase substrate is added. The second test will be a verification to see if the modified plasmid contains the expected restriction enzyme sites.

Precautions

Follow standard laboratory safety procedures. Do not directly contact any chemicals and discard all materials that have contacted bacteria in the biohazard waste. Follow the precautions for agarose gel electrophoresis described in Laboratory 2.

Procedures (See Figure 17.1):

A. Detecting Luciferase Gene Expression

1. Inoculate 2–5 separate colonies from your transformed plates into 5 ml of LB broth with 150 µg/ml ampicillin **the day before this laboratory activity**. Also inoculate 1 tube from your stock culture of cells containing pGEM®-*luc* or from a

Materials

- Luciferase assay system with reporter lysis buffer (Promega Cat #E4030)
- Plasmid preparation kit
- Sterile test tubes with 5 ml of LB with ampicillin
- Sterile 20% glycerol

Photographic Atlas Reference
Chapter 9

white colony on the pGEM®-11zf, pGEM®-*luc* transformation in Lab 13. Label your tubes with your initials and "luc" for the tube containing cells with pGEM®-*luc*, "link" for the tubes containing cells with pGEM®-*luc* with a linker inserted. Incubate overnight in a shaking incubator at 37°C

2. The next day, label the appropriate number of microcentrifuge tubes luc-1, luc-2, etc.. Pour 1 ml of the appropriately numbered culture into the appropriately numbered 1.5 ml microcentrifuge tube. Pellet the cells by centrifuging at maximum speed in the microcentrifuge for 2 minutes. Repeat for the cells containing pGEM®-*luc* without the linker.

3. Pour off the supernatant — you want the pellet to be almost dry, so you need to work at it a bit. Add 50 µl of 1X lysis buffer solution—(10 µl 5X lysis buffer, 40 µl distilled or deionized water). DO NOT MIX. Add 100 µl of luciferase assay substrate. DO NOT MIX. Go into a completely dark room. An interior room with no windows works well if it can be darkened completely. Turn off the lights, allow your eyes to adapt to the dark (at least 30 seconds). *Then*, mix cells and solution by flicking with your fingers vigorously. You should see an immediate glow that slowly decays if the bacteria contain the modified plasmid, and no glow in the cells containing the pGEM®-*luc* plasmid. (The light intensity is fairly low; often you can see it most clearly out of your peripheral vision.)

B. Plasmid Preparation

Each person should do plasmid preparations on one or two clones that appear to express luciferase, along with

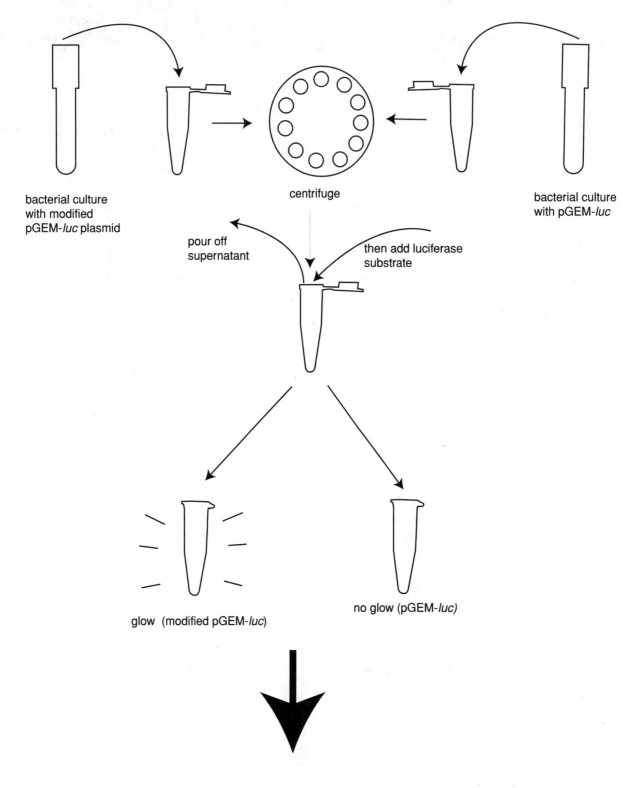

bacterial culture
with modified
pGEM-*luc* plasmid

centrifuge

bacterial culture
with pGEM-*luc*

pour off
supernatant

then add luciferase
substrate

glow (modified pGEM-*luc*)

no glow (pGEM-*luc*)

plasmid preparation, digestion, analysis

Figure 17.1   Overall procedure for Lab 17.

the cells that contain pGEM®-*luc*, as described in Laboratory 14. If you have cells that express luciferase, prepare 1 ml of cells for storage by spinning the cells down, pouring off the supernatant, then resuspending the cells in 20% glycerol. Put the cell suspensions (in glycerol) in the −80°C freezer (−20°C OK for short-term (a few months) storage). Repeat the process for cells containing the original pGEM-*luc* plasmid. These frozen cells may be used in a subsequent laboratory exercise (Labs 21 and 22).

C. Analysis of Plasmids

Clones with a linker inserted should have an extra *Xho*I site in them. When digested with *Xho*I, these plasmids should produce DNA fragments of approximately 1.7 kb and 3.2 kb. The pGEM®-*luc* plasmid itself should only produce a 4.9 kb band.

Set up three restriction enzyme digests. Two will be of your putative luciferase-expressing clones, one your original pGEM®-*luc* plasmid preparation. Add 8 μl of the appropriate DNA solution to each tube. Then add 1 μl of *Xho*I buffer and 1 μl *Xho*I enzyme to each tube. Incubate for at least one hour or overnight at 37°C.

Next Class Period

Run the DNA samples on a 0.8% agarose-0.5 X TBE gel containing ethidium bromide.

Isolating Clones that Express Luciferase

Name:_____   **Date** _____

Results

Questions

1. Why is it essential to wait 30 seconds once you enter a darkened room before you mix the cells and the luciferase substrate?

2. The fact that firefly luciferase can be expressed as an active protein in *E. coli* implies what about the genetic code?

3. Firefly luciferase is commonly used as a reporter gene. A reporter gene is a gene which can be coupled to a promoter and used to monitor when a particular promoter is active. Why is luciferase a good reporter gene?

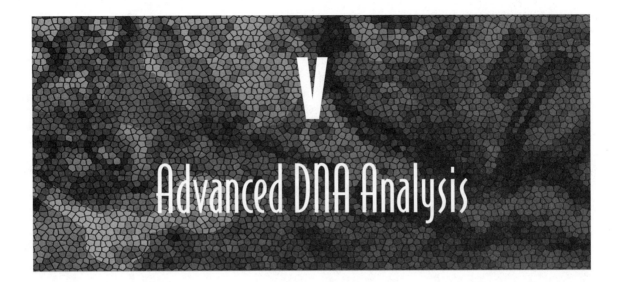

V
Advanced DNA Analysis

The exercises in this section of the lab manual are designed to teach you several additional methods for analyzing DNA. In Lab 18 you will sequence the DNA you mutated in Lab Exercises 13–17 (or another DNA as described by your instructor). In Lab 19, you will use the computer to analyze sequence data. In Lab 20, you will then learn a common technique for studying protein-DNA interactions — the Gel retardation assay.

Key Terms

DNA sequencing: a method for determining the exact order of bases in a DNA molecule

ddNTPs: dideoxy nucleotide triphosphates; nucleotides that lack a 3'-OH group, and terminate DNA synthesis. Termination of DNA chains by ddNTPs produces a ladder of bands used to determine the DNA sequence.

DNA sequencing has become an invaluable technique in molecular biology. One can verify plasmid constructions and compare gene sequences of unknown function with those of known function to determine their possible role in the cell. In the case of genome projects, sequencing allows one to catalog the complete set of genes required to specify a particular organism.

Until recently, most DNA sequencing was done manually, using radioactive nucleotides for detection. Currently, by volume, much more DNA is being sequenced by automated DNA sequencing apparati, which analyze DNA in which fluorescent dyes have been incorporated. This laboratory exercise is a hybrid: you will use a non-radioactive method (digoxigenin labeling and detection) to identify DNA run in

Photographic Atlas Reference
Chapter 11

a traditional (manual) way. The overall sequence of laboratory exercises is as follows (see Figure 18.1):

 A. Sequence reactions

 B. Plate preparation

 C. Gel pouring

 D. Running sequence reactions on the gel

 E. Transferring DNA to a membrane; Staining and analyzing the sequence reactions

Precautions

Individual precautions are listed for each section of the protocol. As a general comment, though, be especially careful to follow good laboratory practices in this exercise since some of the reagents and equipment have potential to be hazardous if mishandled.

A. Sequencing Reactions:

These instructions are adapted from the manual accompanying the Dig Taq DNA Sequencing Kit for standard and cycle sequencing (Boehringer Mannheim). Follow the manufacturer's instructions if you are using another kit. If you modified the pGEM®-*luc* plasmid with a linker in the previous lab period, you may wish to sequence that DNA. Otherwise, any purified plasmid DNA, which contains binding sites for the pUC/M13 forward or reverse primers, can be used in the sequencing reaction.

Procedures

1. Label one, 1.7 ml tube with your initials. Obtain four, 0.2 ml or 0.5 ml PCR tubes (depending on the thermocycler you will use). Label all the PCR tubes with your initials. Label one of those tubes

Materials for Section A

- One, 0.5 ml or 1.7 ml microcentrifuge tube
- Four, 0.5 or 0.2 ml PCR tubes
- Mineral oil
- Dig Taq DNA sequencing kit for standard and cycle sequencing; Cat. #1449 443 from Boehringer Mannheim)
- Thermocycler
- Purified plasmid DNA (e.g. the luc linker plasmid from Lab 17)

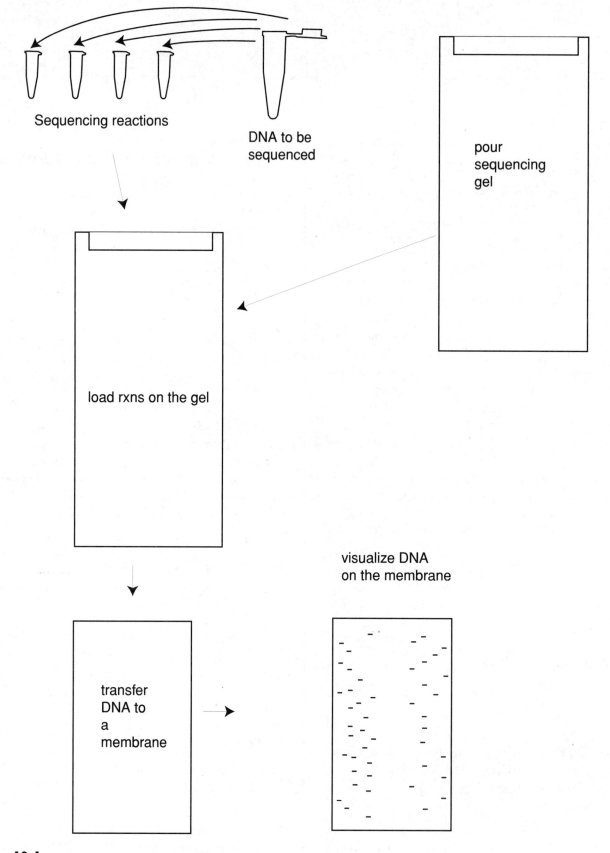

Sequencing reactions

DNA to be
sequenced

pour
sequencing
gel

load rxns on the gel

visualize DNA
on the membrane

transfer
DNA to
a
membrane

Figure 18.1 Overall procedure for Lab 18.

"A", one tube "T", one tube "C" and one tube "G". As an additional precaution, it may be useful to color-code the tubes, by putting a spot of color representing each nucleotide on the appropriate tube. For example, use green for G, red for C, purple for A, black for T.

2. Add 2μl of the dideoxy termination mixes to the appropriate tube. For example, add 2 μl of d/ddATP mix to the A tube, 2 μl of the d/ddTTP mix to the T tube, etc. Place the tubes on ice.

3. To the 1.7 ml tube, add:
   - 10 μl Plasmid DNA (~3 μg, ~ 1 pmole; modified pGEM®-*luc* or other plasmid; for a control reaction, use the appropriate volume of plasmid as directed by the manufacturer)
   - 2 μl DNA reaction buffer
   - 6 μl distilled or deionized water
   - 1 μl dig-labeled M13/pUC (forward) sequencing primer (1 pmole)
   - 19 μl total volume

   Then add 1 μl of Sequencing grade *Taq* polymerase to the tube. Mix by gently pipetting up and down.

4. Add 4 μl of the mix created in step 3 to each of the 4 PCR tubes set up in Step 2. Mix by pipetting up and down. Add a drop of mineral oil to each tube. (Because of the small volume of reagents used in this reaction, it is best to add mineral oil, even if you have a heated thermocycler.) Place the tubes on ice.

5. Place in a preheated (95°C) thermocycler. Heat at 95°C for 2 minutes. Then run samples for 30 cycles of: 95°C for 30 seconds, 60°C for 30 sec., 70°C for 1 minute. The reactions can be left in the machine overnight, if it is set to go to 4°C after 30 cycles.

6. After tubes are removed from the thermocycler, add 2 μl of Stop solution. Store the tubes in the -20°C freezer until needed.

B. and C. Sequencing Gel Preparation/Pouring

Precautions

Acrylamide is a potent, cumulative neurotoxin; *gloves, lab coats and goggles should be worn at all times* when handling unpolymerized acrylamide solutions.

SigmaCote™ and bind silane are toxins and should be handled in a fume hood with gloves.

Glass Plate Preparation

Clean both plates by rubbing vigorously with a Kimwipe in a solution of Liqui-Nox. Rinse plates thoroughly in distilled (RO) water, then deionized water.

Long Plate:

Be sure to handle bind silane in a fume hood with gloves.

Prepare binding solution by adding 3 μl of bind silane to 1 ml of 95% ethanol, 5% acetic acid. Wipe this solution over the cleaned surface of the long plate with Kimwipes®. Allow this solution to dry for 5 minutes.

Then apply 2 ml of 95% ethanol to the plate. Wipe with gentle pressure using a Kimwipe®. Repeat this wash three times.

Materials for Sections B and C
- Glass plates
- Gel spacers
- Gel comb
- Liqui-Nox detergent
- SigmaCote™
- Bind silane, 95% ethanol, acetic acid
- Acrylamide solution (19:1 Acrylamide:bis-acrylamide; a pre-weighed mixture (Sigma Cat #2917) is convenient and reduces exposure to hazardous acrylamide powder)
- Urea
- 5X TBE (prepared previously or as described in the Appendix)
- Distilled or deionized water
- Ammonium persulfate
- TEMED
- Otter™ Casting Stand
- Large paper binder clamps (for holding the glass plates together)
- 50°C water bath
- High voltage power supply
- Sequencing electrophoresis apparatus
- Narrow gauge needle and syringe

Short (eared) Plate:

NOTE: *SigmaCote™ is a toxin and should be handled in a fume hood with gloves!*

1. Put some SigmaCote™ on a Kimwipe®. Wipe the SigmaCote™ over the surface of the plate. Allow to dry 5 minutes.
2. Wipe the surface of the plate with a Kimwipe® to remove excess SigmaCote™.

Gel Preparation, Gel Pouring

Gels will be poured in the Otter™ casting stand. Put paper towels or absorbent bench paper under the top and bottom ends of the plates to catch drips.

Put the bottom (un-notched plate) on the casting stand, clean side of the plate up, and with the top edge of the plate against the far end of the stand. Put the spacers in place, with a few drops of water on each spacer.

Put other glass plate on top, clean side down. Check to make sure the plates are level. If necessary, adjust the level by putting paper towels under the appropriate ends of the apparatus. Once level, slide the top plate back so it overlaps bottom plate by one inch.

Make sure the comb is ready and binder clamps are available. Cut two pieces of plastic wrap to fit over the two ends of the gel.

Preparing a 4% polyacrylamide, 8 M urea gel.

Follow precautions for handling unpolymerized acrylamide.

- Mix together, in a 100 ml, orange-topped bottle:
- 28.4 ml dH2O
- 10 ml 40% acrylamide (19:1 ratio of acrylamide:-bisacrylamide)
- 20 ml 5X TBE
- 48 g Urea

Heat in a 50°C water bath, mixing as you go. It is important that the solution only gets warm enough to dissolve the urea. If it gets too hot, the acrylamide will polymerize too rapidly.

MAKE SURE GEL APPARATUS IS READY!!!
(SEE ABOVE)

Then add 34 *mg* of Ammonium persulfate, swirl to mix/dissolve.

Finally, add 34 µl of TEMED, swirl, start to pour as you slide the upper, notched plate forward. Make sure that the leading edge of the notched plate is always covered with acrylamide. If bubbles form, slide plate back to remove them. When top plate is as far as it will go, clamp the sides with binder clips, slide in comb—*tooth side out, to the appropriate depth*), put plastic wrap over both ends, Place binder clips over the comb (do not place clips on the bottom of the gel). Allow the gel to polymerize overnight (up to two days is normally OK).

D. Gel Running

Precautions

DNA sequencing requires very high voltages. Be sure you know how to properly use the electrophoresis apparatus to reduce any danger of electrical shock or damage to equipment. Regardless of the type of apparatus you have, be sure that the electrodes are disconnected from the power supply and that the power supply is turned off **before** touching the apparatus.

Procedure

With the gel still lying on the casting stand, gently pull out the comb. Squirt distilled water into the gap where the comb had been. Wick out water with a paper towel and reinsert the cleaned comb, this time with teeth pointing down. The teeth of the comb should stick 1–2 mm into the gel. Wipe the dried polyacrylamide off the rest of the gel with a wet paper towel. Remove the binder clips.

Place the gel on the electrophoresis apparatus. Fill the buffer chambers with 1 X TBE. Add sufficient 1X TBE to ensure that both ends of the gel are covered with liquid. Remove bubbles from the bottom of the gel by squirting 1X TBE between the glass plates with a bent needle and syringe. Remove bubbles between the teeth in the top part of the gel by squirting 1X TBE with a narrow-gauge, straight needle and syringe. After the bubbles are removed, load 2–3 µl of stop solution in 5–6 wells spaced across the gel to test the integrity of the wells. Pre-run the gel for 20–30 min at 50W constant power (probably about 1400–1600 V).

While gel is pre-running, heat the sequencing reactions at 95°C for 2 min., then put on ice. After the gel pre-run is completed, turn off the power supply and disconnect the leads. Squirt out urea from the wells with the narrow gauge needle and syringe used to remove bubbles. Then load the gel with about 4 µl per lane of the sequencing reactions. Normally, the gel is loaded TCGA. Run the gel for 2 to 2.5 hours. The run can be stopped when the lower, dark blue dye (bromophenol blue) is within 5 cm of the bottom of the gel. When the gel has finished running, follow the protocol for transfer and detection (below).

Tips

Specially designed, flattened pipet tips can make loading the wells easier, although they are not necessary.

Leaky wells can normally be avoided by using a properly matched set of spacers and gel comb. If you notice that some of the wells leak after the stop solution has been added prior to pre-running the gel, it sometimes helps to push the comb into the gel a little further. If that is unsuccessful, avoid loading samples in the leaky lanes.

E. DNA Transfer, DNA Detection

Procedure

1. Separate the plates: The gel should stick to the long glass plate. Place the gel with the short plate up. Pull out the spacer on one side of the gel. Carefully stick a thin chemical spatula into the bottom corner of the gel and gently twist to pry apart the plates. If the gel sticks to the top plate, try separating the plates very slowly; if that doesn't work, try squirting the gel off the top glass plate with distilled water from a squirt bottle.

2. Cut the membrane to fit the section of the gel that contains DNA samples (you may need to cut several separate sheets, depending on the size of your

---

Materials for Section E

- 1 X TBE buffer
- Nylon membrane (Positively charged; Boehringer Mannheim)
- Washing buffer, Blocking solution, Detection buffer, and Color Substrate Solution (prepared as described in the Appendix).

---

membrane). Mark the membrane with your name at the top with pencil. Always handle the membrane by the edge, with forceps. Place the dry membrane on the part of the gel with your samples (pencil side up). Place two sheets of dry Whatman paper on top of the gel. Then place a sequencing gel plate on top of the paper, and about 3 kg (6 thick books) on top of the plate. Allow the transfer to occur for 30 minutes.

3. Remove your membrane from the gel (gently wash off any gel that may be sticking with 1X TBE buffer). Place the membrane, DNA side down, on plastic food wrap on the UV transilluminator. Irradiate for 3 minutes.

All the following steps are done at room temperature with gentle shaking (except Step 8, which should be done without shaking).

4. Place the membrane into a large plastic, sealable bag (for long, narrow membranes, clean petri dish bags work well). Add 0.10 ml blocking solution per square centimeter of membrane (e.g. for a 20 cm X 30 cm membrane, add 60 ml of blocking solution). Squeeze out air bubbles, seal the bag, and agitate for 30 min.

5. Open the bag, pour off Blocking buffer, add fresh Blocking buffer containing anti-digoxigenin antibody (0.1 ml per square centimeter of membrane, with a 1:10,000 dilution of antibody; for a 20 X 30 cm membrane, add 60 ml of blocking buffer containing 6 µl of anti-digoxigenin antibody). Seal the bag and incubate at room temperature for 15-30 min (this incubation can be extended to at least two hours without causing problems).

6. Open the bag, discard the blocking solution containing the antibody. Add Washing buffer (0.2 ml per square centimeter of membrane; for a 20 X 30 cm membrane add 120 ml of washing buffer). Reseal the bag, agitate for 10 min. Cut open the bag, discard the washing solution, add 120 ml of fresh washing buffer, reseal the bag, incubate for another 10 minutes.

7. Open the bag, pour off the Washing buffer, add 30 ml of Detection Solution, incubate for 2 minutes.

8. Pour off the Detection solution, then add just enough Color Substrate solution or Western Blue® (Promega Biotech) to cover membrane (0.05 ml per square centimeter of the membrane). Incubate

without shaking until color appears (make sure membrane is DNA-side up). Normally the blot can sit in the Color Substrate solution overnight (in the dark) in a sealed bag.

9. When color is sufficiently dark (typically after no more than 16–18 hours), pour off detection solution, rinse with distilled or deionized water or TE. Allow blot to dry in your drawer (i.e., in the dark) on a paper towel.

References

The DIG System User's Guide for Filter Hybridization. 1995. Boehringer Mannheim, Indianapolis, IN

Sambrook, J. E. Fritsch, T. Maniatis. 1989. Molecular Cloning. A Laboratory Manual. 2nd Ed. Cold Spring Harbor Laboratory Press; Cold Spring Harbor, NY.

DNA Sequencing

Name:_____   Date _____

Results

Questions

1. Why was it important that the DNA primers carry a single digoxigenin label, compared to multiple, varied numbers of digoxigenin molecules on the DNA primers?

2. What would happen if you omitted dideoxynucleotides from the sequencing reaction mixture?

3. In traditional DNA sequencing methods (ones that don't use a thermocycler) the DNA must be either single stranded or denatured prior to starting the procedure. That is not required in this case. Explain why.

Computer Analysis of DNA Sequence Information

DNA sequence databases are growing at a rapid rate. Computers are now used for most manipulation and analysis of DNA sequences. There are many programs for DNA sequence analysis available commercially; there are also a number of very useful free web sites and tools for DNA sequence analysis.

In this exercise, you will be asked to do three things:

1. Input the DNA sequence that you determined in the previous laboratory session, and look for "hits" — similar sequences in the databases.

2. Try to determine the function of an "unknown" sequence by inputting it into a database.

3. Analyze the genome of an organism to see how much progress has been made in this field over the last several years.

Looking for Database "Hits"

1. Open your computer's internet browser and point it at the following URL:

    http://www.ncbi.nlm.nih.gov/BLAST/

   - Click on "basic blast search"
   - Enter a 25 to 30 nucleotide sequence from your sequencing gel into the box in your web browser. If your sequencing reaction was not successful, enter the sequence below (it is from a section of the luciferase gene from *Photinus pyralis*).

        TCTTCCAGGGATACGACAAGGATATGGG

Materials

- Computer with internet access, Internet Browser software
- Manually recorded DNA sequence from the previous laboratory session

---

- Hit the "search" button. A screen will come up posting a description of how long it will take for the search to be completed.

- Hit the "Format results" button when you are ready to see your results.

You will be given a results list. Interpret the matches based on the "E" values given. The "E" value indicates the expected number of sequences in a database that would achieve a match as high as the observed value by chance. "E" values below 0.01 would indicate that the sequence match identified would occur only rarely by chance. "E" values of 1e - 10 mean that a random match as good as that observed would be expected to occur only one in 10 billion times. Therefore, a very low E value means that match is very good.

- Answer the questions in the worksheet.

2. Identifying the probable function of a gene by computer matching.

   - Go back to the BLAST home page and choose blastp ("p" for protein) program for the search (the default is blastn).

   - Enter the following sequence:

     GTRAVDGLDLNVPAGLVYGILGPNGAGKSTTIRMLATLL

   - Hit the "Search" button.

   - On the next screen, hit the "Format results" button.

   - Answer the questions in the worksheet.

3. Genome analysis. If you have time, you may wish to peruse several WWW sites that have sequence information of the whole genomes of organisms. Some web sites you may wish to try are:

   - TIGR; a number of microbial genomes, as well as partial parasite and human genomes (as of 1999).

        http://www.tigr.org/tdb

- The National Center for Biotechnology; links to genome work on humans, mouse, and other organisms

  http://www.ncbi.nlm.nih.gov/

- The Sanger Center; information on a number of DNA sequencing projects

  http://www.sanger.ac.uk/

Reference

Trends Guide to Bioinformatics. 1998. Elsevier Science, Cambridge, UK.

Computer Analysis of DNA Sequence Information

Name:_____   **Date** _____

Results

Questions

1. If you entered the DNA sequence given in the exercise, you should have gotten a number of sequences with identical, very low E values. Explain this result.

2. The DNA sequence given in the exercise was 28 nucleotides long and should have produced very small E-values. When an 18 nucleotide sequence of the same gene was used as a query for a blast search, the smallest E-value produced was 0.011. How do these two sets of E-values help explain why 18-mer or longer oligonucleotides are typically used for PCR when eukaryotic genomic DNA is used as a template?

3. When you entered the amino acid sequence given in the exercise, what sequence did that sequence match? Explain your results.

4. Based on the matches to other proteins, what can you infer about the function of the protein? Explain why this type of sequence analysis could be useful for further experimentation involving determining the function of a gene?

Key Terms

Gel retardation Assay: a method for determining whether a protein binds to a particular DNA sequence. The method is based on the slower migration of a DNA-protein complex through a gel, compared to the migration rate of DNA alone.

Gel retardation assays are designed to detect binding of a specific protein to a specific DNA sequence. In this gel retardation experiment, you will be looking at the specific binding of a protein (antibody) to digoxigenin-labeled DNA.

Precautions

Follow the standard precautions for preparing and analyzing ethidium-bromide stained agarose gels.

Materials

- Dig-labeled *PGK* PCR product (from Northern blotting laboratory)
- Non-dig-labeled *PGK* PCR product (from RT-PCR laboratory or follow the protocol for preparing a probe for the Northern blotting laboratory, except use 1 µl of 10 mM dNTP mix instead of the labeling mix with digoxigenin).
- 10X Maleic Acid Buffer (recipe in Appendix; used previously in Southern blotting and other labs)
- 4 microcentrifuge tubes
- loading dye
- Agarose, 0.5X TBE buffer, electrophoresis apparatus.
- p20 pipettors, pipet tips
- Anti-digoxigenin antibody

Photographic Atlas Reference
Chapter 12

Procedures (See Figure 20.1)

1. You will need purified dig-labeled, and unlabeled *PGK* PCR products. If you did the Northern blotting laboratory, you should have extra, purified, dig-labeled PCR product that you can use in this exercise. If you did the RT-PCR exercise, you should have unlabeled *PGK* PCR product. Purify this product by electrophoresis through an agarose gel followed by purification through a spin column (Lab 5) or by gel filtration of the PCR product (Lab 16).

   If you did not do those laboratory exercises, follow the procedures for generating the PCR products as given in the northern hybridization laboratory exercise (Lab 12). In this case, do two reactions; one exactly as given in the exercise (with dig-dNTP labeling mix) and the other reaction with 1 µl of 10 mM dNTPs and 4 additional µl of water.

2. Pour a 2.5%, 0.5X TBE agarose gel with ethidium bromide. Prepare the gel as described previously, but be aware that gel solutions that contain this much agarose boil over easily. It may be useful to allow the solution to boil once, swirl, then allow the solution to sit for 5 minutes before continuing to microwave the gel.

3. Mix together the following reagents:

   Tube 1:
   7 µl of purified PCR product (dig-labeled)
   1 µl 10X Maleic Acid buffer
   2 µl distilled water

Tube 2:

7 μl of purified PCR product (dig-labeled)

1 μl 10X Maleic Acid buffer

2 μl anti-dig antibody

Tube 3:

7 μl of purified PCR product (not dig-labeled)

1 μl 10X Maleic Acid buffer

2 μl distilled water

Tube 4:

7 μl of purified PCR product (not dig-labeled)

1 μl 10X Maleic Acid buffer

2 μl anti-dig antibody

Allow the tubes to sit at room temperature for 10 minutes. Then add 2 μl of 6X DNA loading dye, load the samples on the agarose gel prepared in step 1. Run the gel in 0.5X TBE buffer at 120V for about 1 hour. Then analyze the pattern of bands.

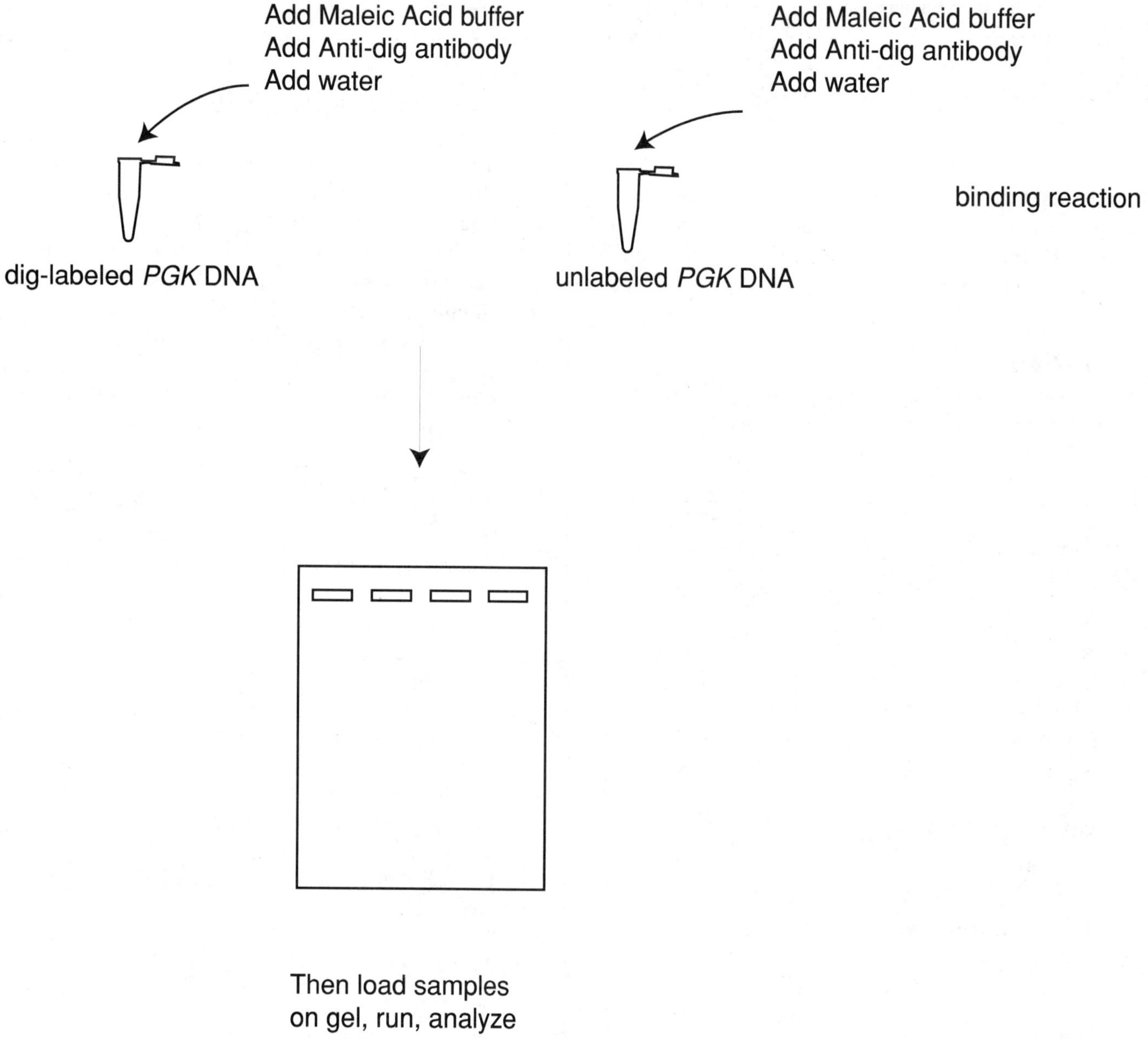

Figure 20.1 Overall procedure for Lab 20.

Gel Retardation Assay

Name:_____   **Date** _____

Results

Questions

1. Would this assay work if you boiled the antibody prior to adding it to the DNA? Why or why not?

2. How would you do an assay for a protein that only bound DNA when it, in turn, bound a small molecule?

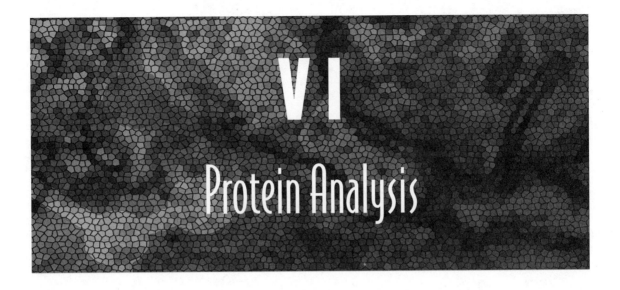

VI
Protein Analysis

Although most of molecular biology focuses on the study of DNA and RNA, protein analysis is also an important tool for Molecular Biologists and Molecular Geneticists. In this section of the laboratory manual, you will learn two common methods for protein analysis; Sodium Dodecyl Sulfate — Polyacrylamide Gel Electrophoresis (SDS-PAGE, Lab 21) and Western Blotting of proteins separated by SDS-PAGE (Lab 22).

SDS-PAGE separates proteins by size; western blotting identifies a protein from an SDS-PAGE gel by the ability of a specific antibody to bind to the protein. These techniques are important for monitoring the expression of proteins in recombinant cells, for protein purification, for the analysis of protein abundance, and for many other purposes.

SDS-PAGE Analysis of Proteins

Sodium Dodecyl Sulfate-Polyacrylamide Gel Electrophoresis (SDS-PAGE) is a commonly used method of separating proteins by size for further analysis or purification. It is a powerful technique that is widely used in molecular biology.

Precautions

Always handle acrylamide solutions with gloves. Unpolymerized acrylamide is a cumulative neurotoxin, and a potential carcinogen and teratogen. Wear eye protection and other laboratory safety equipment as directed by your instructor.

Always avoid contact with chemicals. Many of the other chemicals used in this laboratory are known to

Materials

- Two, 1.7 ml microcentrifuge tubes
- Lid-lock
- 2X Gel loading buffer
- SDS-PAGE apparatus
- 4% agar in distilled water (not agarose)
- 40% Acrylamide solution (29:1 Acrylamide:-Bisacrylamide)
- TEMED
- Ammonium persulfate
- 4X Resolving Gel buffer
- 4X Stacking Gel buffer
- 20% SDS
- 5X Tris-Glycine Running buffer
- Protein markers
- Coomassie stain
- Destaining solution
- *E. coli* overnight culture, 1 ml (with the modified Luciferase plasmid)
- *E. coli* overnight culture, 1 ml (without the modified Luciferase plasmid)

Photographic Atlas Reference
Chapter 13

be harmful if handled improperly (e.g., methanol, acetic acid, TEMED). Wear gloves and other protective equipment as directed by your instructor.

Additional Instructions

- If you are doing the next exercise, you will also need to prepare the transfer buffer described in Lab 23.
- If you are not doing Lab 23, any overnight-grown *E. coli* strain will work.

Preparation of Solutions

4 X Resolving Gel Buffer (1.5 M Tris-HCl, pH 8.8)

Add 36.3 g Tris *base* to 150 ml distilled or deionized water. Adjust pH to 8.8 with HCl. Add deionized or distilled water to a final volume of 200 ml. Autoclave to sterilize.

4 X Stacking Gel Buffer (0.5 M Tris-HCl, pH 6.8)

Add 3.0 g Tris *base* to 40 ml distilled or deionized water. Mix to dissolve. Adjust pH to 6.8 with HCl. Add deionized water to a final volume of 50 ml. Autoclave to sterilize.

20% SDS

Add 20 g of SDS to a final volume of 100 ml of distilled or deionized water.

5X Tris-Glycine Running Buffer

Add 15.1 g Tris base and 72 g of glycine to 900 ml deionized or distilled water. Add 25 ml 20% SDS. Adjust the volume of the solution to 1000 ml.

For the 1X Running buffer, add 100 ml of 5X Running buffer to 400 ml of distilled or deionized water.

Acrylamide

29:1 Acrylamide: Bisacrylamide solution.

Sigma premixed Acrylamide: Bisacrylamide powders, which only require the addition of distilled or deionized water, are recommended.

2X Protein Gel-loading Buffer

(100 mM Tris-HCl, pH 6.8, 200 mM DTT, 4% SDS, 0.2% Bromophenol blue, 10% glycerol)

For 1 ml of Protein gel loading buffer, add
- 0.2 ml of 4X stacking gel buffer
- 0.2 ml distilled or deionized water
- 0.2 ml 20% SDS
- 0.1 ml glycerol
- 0.1 ml 2% Bromophenol blue.

This solution can be stored at room temperature. Just before use, add 1/5 volume (0.2 ml) of 1 M DTT to the loading buffer.

1 M DTT

Dissolve 0.15 g DTT in 1 ml distilled or deionized water. Portion out into 0.2 ml aliquots in microcentrifuge tubes, store at −20°C.

4% Agar

Add 4 gm of agar to a final volume of 100 ml with distilled water.

For Staining

Coomassie Blue Stain (0.06% Coomassie blue G-250)

For 1 liter, add 0.6 gm Coomassie Blue G-250 stain to 900 ml water. Add 100 ml glacial acetic acid, mix.

Destaining solution: (7% acetic acid, 5% methanol)

For 1 liter, add 70 ml acetic acid to 50 ml methanol. Add distilled or deionized water to 1 liter.

Procedures (See Figure 21.1)

Preparing the Glass Plates

Clean the plates thoroughly with detergent, then rinse with distilled water. Clean the spacers and comb in the same way.

Some models of electrophoresis equipment have special gel pouring stands. If so, follow the manufacturer's directions. If not, follow the directions below for pouring the gels.

Using binder clips, assemble the two glass plates with spacers inserted in between. Be sure the bottom binder clip is exactly level with the bottom of the gel (i.e., the gel should be resting on the binder clips and the bottom of the gel). Set assembled gel on top of a clean glass plate.

Place the comb between the glass plates. Measure down 1 cm. Make a mark. This mark represents the boundary between the stacking and resolving gels.

Melt the agar solution in the microwave (watch carefully, swirl frequently to prevent boil-over).

Pour a small amount of agar on the glass plate. Set the gel apparatus into the melted agar. The agar should wick up between the glass plates. Allow the agar to solidify. The agar should plug the bottom end of the gel set-up.

Pouring the Gel

Resolving Gel (8% gel, 30 ml). Adjust reagent volumes for smaller or larger gel sizes

Be sure your glass plates are ready to accept the acrylamide solution (as described above) before you start preparing this solution.

Resolving Gel

Mix together in a flask or beaker:
- 6 ml 40% Acrylamide/Bis (29:1)
- 16 ml distilled or deionized water
- 7.5 ml 4X Resolving gel buffer
- 0.15 ml 20% SDS
- 0.3 ml 10% Ammonium persulfate

Mix by swirling gently. *Then add 19 μl TEMED.* Pipet or pour the gel quickly into the gel mold to the appropriate depth (based on the mark you made on the glass plates). Gently pipet 0.1% SDS to cover the gel. Allow the resolving gel to polymerize for at least 30 min.

Stacking Gel

After the resolving gel has polymerized, pour off the 0.1% SDS, and rinse the gel with distilled water. Dry the gel by inversion, then pick up any extra water between the plates with a paper towel. Then prepare the stacking gel.

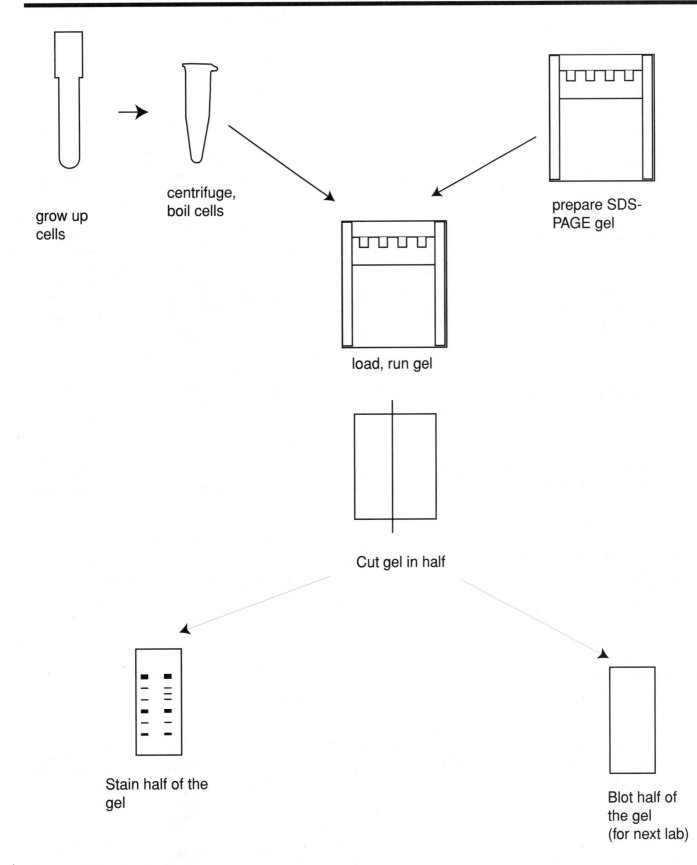

grow up
cells

centrifuge,
boil cells

prepare SDS-
PAGE gel

load, run gel

Cut gel in half

Stain half of the
gel

Blot half of
the gel
(for next lab)

Figure 21.1 Overall procedure for Lab 21.

For a 5 ml stacking gel, mix together:

- 0.62 ml of 40% Acrylamide:Bisacrylamide (29:1)
- 3.1 ml distilled or deionized water
- 1.25 ml of 4X stacking buffer
- 25 µl 20% SDS
- 50 µl 10% Ammonium persulfate

Swirl to mix. Then add 5 µl TEMED, and pipet or pour the acrylamide solution into the gel. Insert the appropriate comb, cover the top of the gel with plastic wrap. Allow the gel to polymerize (~30 min.).

Luciferase-expressing Strain

If you are planning to identify the luciferase protein, you should have two separate cultures. Once culture should contain the pGEM®-*luc* plasmid you modified, the other should contain either unmodified pGEM®-*luc* or pGEM®-llzf.

If you are not planning to do laboratory 23, then simply grow up a single *E. coli* culture.

Preparation of the Cell Lysate

While the stacking gel is polymerizing, spin down two, 1 ml bacterial cell cultures. Pour off the supernatant, resuspend the pellet by vortex mixing. Add 10 µl of distilled water. Vortex mix. Add 10 µl of 2X Protein gel loading buffer containing DTT. Place a lid-lock on the tubes. Boil the samples for 3 min. Centrifuge the samples for 3 min. Place the tubes on ice.

Loading and Electrophoresis of Samples

Remove the comb from the gel apparatus. Using a squirt bottle, wash the wells with deionized water. Put the gel in the electrophoresis apparatus. Add the appropriate amount of 1X running buffer to the top and bottom buffer chambers. Remove bubbles from the bottom of the gel and from the wells with a syringe by squirting buffer into the wells.

Load the pre-stained protein markers in the middle well of the gel.

Load up to 10 µl of your protein sample per well. Load duplicate samples on either side of the protein markers if you are continuing with lab 23. Otherwise, load the one sample per gel.

Carefully pipet in your sample so it does not flow over into the next well. When everyone has loaded their samples, attach the electrodes to the power supply (positive electrode to the bottom buffer chamber). Run the gel at 8V/cm of gel length. Once the dye has moved into the resolving gel, turn the voltage up to 15V/cm of gel length. When the bromophenol blue dye reaches the bottom of the gel, turn off the power.

Staining of Proteins (From Hoefer Protocol)

Be sure the power is off. Dump out the buffer from the top chamber. Remove the gel from the electrophoresis apparatus. Gently pry the plates apart with a thin chemical spatula. Cut or pull off the stacking gel. If you are continuing with lab 23, cut the gel in half. One half of the gel will be stained, the other half transferred to a membrane. Otherwise, leave the gel intact if you are only staining.

Remove half (or all) of the gel from the plate and gently transfer to a staining container with a tight-fitting lid (plastic food containers work well). Add sufficient Coomassie stain to completely cover the gel. Allow the gel to sit in stain for at least two hours or overnight.

Pour off the Coomassie stain, add destaining solution, incubate until the background becomes clear. Slow shaking speeds destaining, as does occasional changes of destaining solution.

Generally, protein bands are most easily visible when viewed on a white light table.

Transfer of Proteins to a Membrane

If you are doing the western blotting laboratory exercise, transfer the proteins from the other part of the gel to a PVDF membrane as directed in the following laboratory exercise. This can be done while the other half of the gel is staining.

References

Protein Electrophoresis Applications Guide. 1994. Hoefer Scientific Instruments, San Francisco, CA.

Sambrook, J., E. Fritsch, T. Maniatis. 1989. Molecular Cloning. A Laboratory Manual. 2nd Ed. Cold Spring Harbor Laboratory Press; Cold Spring Harbor, NY.

SDS-PAGE Analysis of Proteins

Name:_____   Date _____

Results

Questions

1. What would happen to positively charged proteins during electrophoresis if no SDS was added?

2. What could explain the migration of a protein in a gel at 100,000 daltons, when DNA sequencing indicates the gene should encode a protein at 60,000 daltons?

3. Describe how these gels could be used for protein purification.

Western blotting is done after SDS-PAGE analysis. It allows for the identification of a specific protein, based on the ability of an antibody to bind to that protein. Western blotting requires transfer of proteins from the polyacrylamide gel to a membrane. This is followed by incubation with an antibody specific for the protein of interest and detection of the bound antibody.

Precautions

■ The transfer buffer used in this exercise contains methanol. **Wear appropriate protective clothing and use appropriate ventilation to prevent inhalation or absorption of methanol.**

■ If working with powdered SDS be sure to wear an appropriate dust mask and wipe up any spilled powder.

■ The blotting apparatus used in this lab can cause serious electrical shock if not handled correctly. Be sure to follow the manufacturer's suggested procedure for using the blotting apparatus.

Materials
■ Polyacrylamide gel with separated proteins from the previous exercise
■ Immobilon-P membrane (Millipore)
■ Blotting paper (e.g., Whatman 3MM paper)
■ Transfer Buffer
■ Electroblotting transfer apparatus
■ Anti-luciferase antibody (Sigma Cat #L0159)
■ Western Blue® Alkaline Phosphatase Substrate (Promega Biotech)
■ Anti-rabbit antibody, conjugated to Alkaline Phosphatase (Promega Biotech)
■ p20 pipettor, pipet tips
■ Phosphate Buffered Saline (PBS)
■ Bovine Serum albumin
■ Tween® 20

Photographic Atlas Reference
Chapter 13

PBS (Phosphate Buffered Saline)

Per liter: 0.23 g NaH_2PO_4 (anhydrous monobasic sodium phosphate), 1.15g Na_2HPO_4 (dibasic sodium phosphate), distilled water to 1 liter. Autoclave to sterilize at 15 lbs. pressure for 15 min. (Also available as a pre-packaged powder from Sigma Chemical Company.)

Blocking Buffer (1% BSA in PBS with 0.05% Tween® 20)

Per 100 ml: Add 1 g Bovine Serum albumin, and 50 µl Tween®-20 to 100 ml PBS.

Transfer buffer (25 mM Tris, 192 mM Glycine, 10% Methanol, 0.1% SDS)

For 100 ml, add 0.3 g Tris base to 70 ml distilled or deionized water. Then add 1.4 g glycine, 0.1 g SDS. Add 10 ml methanol. Adjust volume to 100 ml.

Procedures (See Figure 22.1)

Transfer of proteins to PVDF membrane

The directions below are for transfer with an Alltech semi-dry blotting system. If you plan to use a tank electroblotting apparatus, or a different type of semi-dry blotter, follow the specific instructions from the manufacturer.

Cut 6 pieces of blotting paper slightly smaller than the dimensions of the gel.

Cut one piece of Millipore Immobilon-P PVDF membrane slightly smaller than the gel.

NOTE: Always handle the membrane with forceps and by the edges.

1. Equilbrate the gel in transfer buffer for 15 minutes.

2. While the gel is equilibrating, soak the membrane in 100% methanol for 15 seconds, then transfer the membrane to distilled or deionized water for two minutes. Finally, place the membrane in transfer buffer until you are ready to set up the transfer stack.

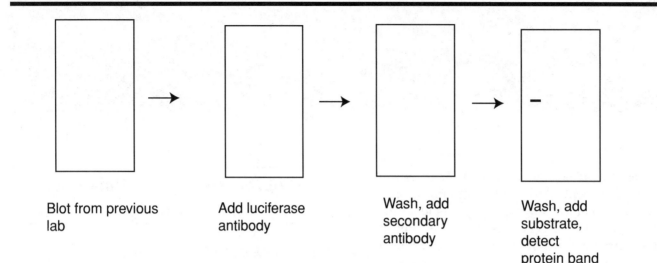

Blot from previous lab

Add luciferase antibody

Wash, add secondary antibody

Wash, add substrate, detect protein band

Figure 22.1 Overall procedure for Lab 22.

3. Soak a piece of blotting paper in transfer buffer. Place the wet paper on the bottom (anode) of the apparatus. Roll any bubbles out with a pipet. Repeat the process with two more pieces of blotting paper. You should now have a stack of three bubble-free sheets of blotting paper, soaked in transfer buffer, on the bottom electrode.

4. Using forceps, carefully place the membrane on top of the stack. Then, carefully lay the gel on top of the membrane. It is critical that the blotting paper and the membrane do not extend past the gel.

5. Then place three pieces of blotting paper that are soaked in transfer buffer on top of the gel. Lay the sheets on one at a time, rolling out bubbles after each sheet is applied.

6. Place the top (cathode) plate on top of the stack. Do this carefully, so as not to move the stack. Connect the electrodes to the power supply. Apply the appropriate current as recommended by the manufacturer of the apparatus. For the Alltech apparatus, protein transfer is normally complete in 30 minutes at 200 mA.

7. After the appropriate length of time, turn off the power supply, disconnect the leads, and remove the membrane from the transfer apparatus with forceps.

If the pre-stained proteins are no longer present in the gel, but are visible on the membrane, this indicates that transfer was successful.

Detecting a Specific Protein on the Membrane

Thoroughly dry the membrane. Place in an air incubator at 37°C for a full hour after transfer or allow it to dry on the bench for at least two hours. Allowing the membrane to air-dry for at least a week at room temperature does not adversely affect the ability to detect the luciferase protein.

Detecting Luciferase

Incubate the blot for 30 minutes with a 1:5000 dilution of anti-luciferase antibody in blocking buffer. Pour off the solution, wash 2 times for 10 seconds each in PBS.

Incubate with anti-rabbit antibody conjugated to alkaline phosphatase for 20 min. The antibody should be diluted 1:5000 in blocking buffer. Pour off the solution, wash the blot 2 times for 10 seconds each time with PBS.

Add sufficient substrate (e.g., Western Blue®) to cover the blot, incubate at room temperature, without shaking, in the dark. When the bands are visible and the background begins to increase, pour off the substrate and wash with distilled or deionized water. (Normally, protein bands take at least 15 to 30 minutes to become visible.) Allow the membrane to dry.

References

Millipore Corporation. Immobilon-P transfer membrane. Millipore, Bedford, MA

Protein Electrophoresis Applications Guide. 1994. Hoefer Scientifc Instruments, San Francisco, CA.

Western Blotting

Name:_____   **Date** _____

Results

Questions

1. You are working on this lab exercise and you run out of anti-luciferase antibody. Could you substitute anti-digoxigenin antibody? Why or why not?

2. How do you know that the proteins transferred out of the gel and on to the membrane?

3. Sometimes you may get additional labeled protein bands when doing western blotting. Provide a possible explanation for these bands.

4. Explain how this technique could be modified to study protein distribution in a cell.

Preparation of Frequently Used Solutions

Agarose Gel Loading Buffer

This solution can either be purchased commercially or can be prepared in the laboratory. These gel loading solutions are typically 6X concentration; add 1μl of gel loading buffer for each 5 μl of DNA solution.

Laboratory Prepared Gel Loading Buffer (10 ml)

- 7 ml distilled or deionized water
- 3 ml glycerol
- 25 mg bromophenol blue
- 25 mg xylene cyanol

5X TBE Buffer (10X Stock)

Add 54.0 g of Tris Base, 27.5 g of Boric acid, 3.72 g of $Na_2EDTA._2H_2O$ to a final volume of 1 liter of distilled water. Mix to dissolve. Reagents should be electrophoresis grade. Store at room temperature. Discard if a precipitate forms on the bottom of the bottle of the 5X solution.

0.5 X TBE Buffer

Dilute 5X TBE buffer 1:10 (e.g., 100 ml 5X TBE buffer into 900 ml distilled water). This buffer solution can typically be reused to run several agarose gels.

TE Buffer (for Diluting DNA)

10 mM Tris-HCl (pH 8)
1 mM EDTA (pH 8)

A. Preparing Stock Solutions for TE

Make a 500 ml solution of 1 M Tris, pH8. Add 60.5 g of Tris Base to 375 ml of distilled or deionized water and adjusting the pH to 8 (this will require approximately 21 ml of concentrated HCl). Add distilled or deionized water to 1 liter. Autoclave to sterilize.

Make 50 ml of 0.5 M EDTA, pH 8. Add 9.3 g of disodium EDTA.$2H_2O$ (ethylenediaminetetraacetate) to 38 ml distilled or deionized water. Mix on a magnetic stirrer. Adjust pH to 8 with NaOH pellets (about 1 g). The EDTA will typically not dissolve completely until the pH approaches 8 (it may take 30 to 60 minutes for the EDTA to go completely into solution). Autoclave to sterilize.

B. Making the Working TE Buffer

Sterilize 98.8 ml of distilled or deionized water. Add 1 ml of sterile 1 M Tris, pH 8, 0.2 ml of sterile 0.5 M EDTA, pH 8.

Ethidium Bromide Stock Solution

Be sure to weigh and dispense the ethidium bromide under a chemical fume hood to prevent inhalation of dust. Always handle ethidium bromide solutions with gloves.

Add 10 mg of ethidium bromide to 2 ml of sterile, distilled water (final concentration 5 mg/ml). Mix thoroughly; it will take some time for the ethidium bromide to dissolve. This is a 10,000 X stock solution. Therefore, you will add 1 μl of this solution to each 10 ml of your agarose gel solution. Pre-made solutions of ethidium bromide can also be purchased commercially.

20X SSC

For 1 liter, dissolve 175 g NaCl and 88 g of sodium citrate in 800 ml of distilled water. Adjust the pH to 7.0 with NaOH. Adjust the volume to 1 liter with distilled water.

Solutions for Probe Testing, Southern Blotting, Northern Blotting, and DNA Sequencing (Labs 6, 8, 12, 18)

Be sure to read the information in the northern blotting procedure for information on DMPC-treating these solutions prior to use in that exercise (Lab 12).

Maleic Acid buffer (10X)
(100 mM maleic acid, 150 mM NaCl, pH 7.5)

For 1 liter of a 10X solution, add 11.6 g maleic acid (Sigma M 0375), 8.8 g NaCl, and 3.5 g NaOH (solid) to 800 ml of distilled water. Mix to dissolve, adjust volume to 1 liter, and adjust pH to 7.5 with NaOH. When the pH gets to ~ 6, use a dilute NaOH solution, and add the NaOH slowly to avoid making the solution too alkaline. Autoclave to sterilize.

Washing Buffer (1X) (10 mM maleic acid, 15 mM NaCl, 0.3% Tween®-20, pH 7.5)

For 1 liter, add 100 ml of 10X Maleic Acid Buffer, 3 ml of Tween 20, distilled or deionized water to 1000 ml.

Blocking Solution (1X)

For 100 ml of blocking solution, add 10 ml of 10X Maleic Acid buffer, 2 g of powdered skim milk, distilled or deionized water to 100 ml. Make sure the powdered milk is dissolved completely. Make this immediately before use; it should not be stored. *For the northern blotting (Lab 12) and the DNA sequencing lab (Lab 18), substitute 1 g of blocking reagent (Boehringer Mannheim Cat #1 096 176) for the 2 g of powdered skim milk. This substitution can also be made for Labs 6 and 8. Microwave the blocking solution at low power to dissolve the blocking reagent, being careful that the solution does not boil.*

Detection Buffer (1X) (10 mM Tris-HCl, 10 mM NaCl, pH 9.5)

For 1 liter, add 1.2 g of Tris base and 0.6 g of NaCl to 800 ml distilled or deionized water. Adjust the pH to 9.5 with HCl. Autoclave. $MgCl_2$ can be added, *after the pH is adjusted*, to a final concentration of 5 mM (1 g $MgCl_2$.6H2O per liter). Addition of $MgCl_2$ can speed the rate of reaction, but it also has the potential to increase the background (non-specific) staining. Detection buffer to which $MgCl_2$ has been added should be filtered through a 0.45 μm filter before use. *For northern blotting (Lab 12) and DNA sequencing (Lab 18) the detection solution should not contain $MgCl_2$.*

Color Substrate Solution

Western Blue®, a prestabilized solution of NBT and BCIP (Promega Biotech, Madison, Wl) is recommended as the color substrate solution. If you prepare your own solutions, no more than 30 minutes before use, add 45 μl of NBT (75 mg/ml nitroblue tetrazolium salt in 70% dimethylformamide) and 35 μl BCIP (50 mg/ml 5-bromo-4-chloro-3-indolyl phosphate in dimethylformamide) to 10 ml of detection buffer.

Wash Solution I (2X SSC, 0.1% SDS)

For a 500 ml solution, add 50 ml 20X SSC, 445 ml distilled water and 5 ml 10% SDS. Be sure to add the SDS last. If the SDS is mixed directly into the 20X SSC solution, it will form an insoluble precipitate.

Wash Solution ll (0.5X SSC, 0.1% SDS)

For a 500 ml solution, add 12.5 ml 20X SSC, 482.5 ml distilled water, 5 ml 10% SDS. As above, be sure to add the SDS last.

Index

Acrylamide, 109, 110, 125, 126, 128

Agarose gel electrophoresis, 1, 9, 11, 13, 15, 51, 65, 66, 69, 85, 99
  precaution, 99
  precautions, 51, 85

Ammonium persulfate, 109, 110, 125, 126, 128

Ampicillin stock solution, 93

ApoC2 gene, 51, 54

Bacteriophage l, 17

BLAST search, 115, 117

Blocking buffer, 35, 36, 45, 72, 111, 131, 132

Calcium chloride solution, 79, 80

cDNA, 15

Color substrate buffer, 45, 46

Competent cell freezing buffer, 79

Competent cell preparation, 79

Coomassie stain, 125, 128

Detection buffer, 34, 36, 45, 73, 111, 136

Digoxigenin, 15, 33, 36, 45, 46, 69, 70, 107, 111, 113, 119, 133

Dimethyl Pyrocarbonate (DMPC), 59

EcoRI, 17, 85, 88, 89

Escherichia coli, 17, 79, 80, 82

Ethidium bromide, 9, 10, 17, 18, 21, 27, 66, 71, 88, 89, 101, 119, 135

Gel loading buffer, 125

Gel retardation assay, 105, 119

Hexanucleotide mix, 33

Hybridization, 21, 25, 33, 36, 39, 40, 41, 45, 63, 71, 75

Hybridization buffer, 69, 71

Klenow enzyme, 33, 34

LB Agar, 80, 82

LB Broth, 80, 82, 85, 96, 99

Luciferase, 77, 93, 95, 99, 101, 103, 125, 128, 131, 132, 133

MacConkey agar, 79, 80, 82, 85, 97

Maleic Acid buffer, 34, 72, 120, 136

Microcentrifuge, 21, 23, 27, 28, 33, 39, 40, 51, 60, 62, 65, 71, 88, 89, 95, 107
  microcentrifuge tubes, 80, 82
  tubes, 17, 35, 54, 59, 79, 80, 99, 119, 125, 126

mRNA isolation, 59, 60, 63, 65

Mutagenesis (linker-based), 93

Nylon membrane, 15, 21, 23, 69, 71

pGEM®-llzf, 128

pGEM®-luc, 77, 82, 89, 93, 94, 95, 99, 109, 128

PGK, 69, 119

Phosphoglycerate kinase gene (PGK), 59, 65, 66

Pipettor, 1, 3, 4

Plasmid, 15, 77, 79, 80, 82, 88, 89, 91, 93, 94, 99, 107, 109, 125, 128

Polyacrylamide, 110, 123, 131

Polymerase Chain Reaction (PCR), 15, 51, 55, 65

Prehybridization, 39, 40, 46, 72
  buffer, 41, 69, 71, 72

Resolving gel buffer, 125, 126

Restriction enzymes, 17

Reverse-Transcriptase-Polymerase Chain Reaction (RT-PCR), 60, 65, 66, 69, 70, 75, 119

RFLP analysis, 15, 17

RNA, 57, 59, 60, 62, 63, 65, 66, 67, 69, 70, 71, 73, 75, 123

RNase, 62, 63, 65, 66, 69, 71, 85

RNase-free, 59, 60, 62, 65, 66, 69, 71

Safety, 99, 125

Sodium Dodecyl Sulfate-Poly-acrylamide Gel Electrophore-sis (SDS-PAGE), 125
  solutions required, 125

Southern transfer, 21, 23

Spectrophotometry, 62

Stacking Gel buffer, 125, 126

Taq polymerase, 65, 69, 109

TBE buffer, 9, 10, 17, 21, 27, 71, 111, 120, 135

TE buffer, 9, 10, 35, 70, 85, 88

TEMED, 109, 110, 125, 126, 128

Tri-Reagent™, 59, 60

Tris-Glycine Running buffer, 125

Urea, 110, 111

UV transilluminator, 9, 11, 23, 35, 111

Vortex mixer, 51, 54, 59

Wash solutions, 45

Washing buffer, 72, 73, 111, 136

Western Blotting, 123, 128, 131, 133